CONTENTS

JANUARY
FEBRUARU
MARCH
2018

03 THE RSM
What is the Rational Scientific Method? What's wrong with the old one?

05 RATIONAL
What is a rational definition? What is a rational explanation? What is a rational person?

08 RSM
The Father of Rational Science, Bill Gaede, explains the Rational Scientific Method

11 PHYSICS
What is Rational Physics? Physics fails to define its Key Term: EXIST

12 TECHNOLOGY
What is the difference between science and technology?

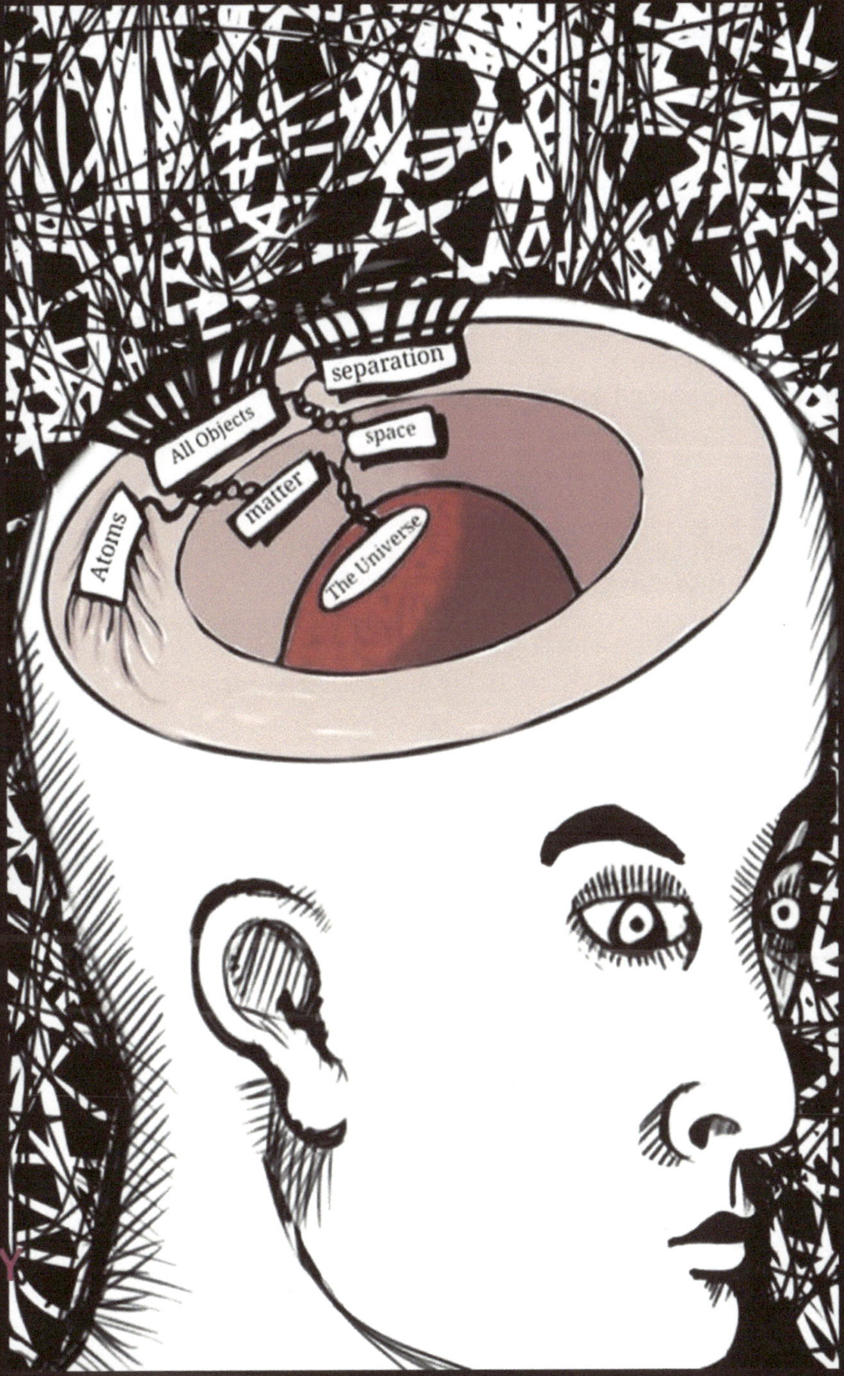

14 EXPERIMENT
Do it on your own time! It's not a part of the RATIONAL Scientific Method

19 INTERVIEW
The MonkE queries Bill Gaede in this exclusive one on one

25 Knowledge
The difference between knowledge and prediction

26 SHADOW
Shadows, reflections, holograms; do they exist?

30 RSM
David Robison pitches his take

6 What Is Rational and What is logical?
17 Word Magic V 1.1, 51 Words Mean Things,

Cover Art by Grinning Monkey Publishing   The Rational Scientist   January, February March

Welcome to the Premiere issue of The Rational Scientist!

A quarterly Magazine

Available on Amazon.com in both E format for immediate download, and 8.5 x 11 paper format.

Our mission is to introduce you to the Rational Scientific Method and to demolish current mainstream nonsense such as Relativity, Quantum Mechanics, String Theory, Big Bang, Black Holes, Faster Than Light, Warped Space, Multi-Dimensions and TimeTravel.

Science magazines have gone astray. Today, the popular rags of science are more interested in scientific fantasy, and political debate or social justice than they are in explaining phenomena using objects. But, then, mainstream science abandoned using objects to mediate phenomena long ago in lieu of abstract theoretical mathematical descriptions and reification.

In this issue we'll get a brief glimpse of what Rational Science means, and with the special interview with Bill Gaede we'll lay the ground work for future issues where we will look at examples of how concepts have been turned into objects and verbs into nouns through the current offerings of Discover, Scientific American, National Geographic, Popular Science and Astronomy magazine. We'll introduce the amazing Rope Hypothesis and Thread Theory and end the year by discussing, with the luxury of detail, The Extinction of Man. Discover the underlying mechanisms behind all extinctions that have ever occurred and understand why man's intelligence and technology have no power to avert this very natural phenomenon.

Editor and Chief
MonkEmind

Executive Editor
MonkEmind

Creative Director
MonkEmind

Contributors
Bill Gaede
David Robison
MonkEmind

Artwork
Daniel Ferguson
David Robison
Mike Huttner
Jeff Hancock

The Rational Scientist
Published by

monkEmind@gmail.com

Cover Art
The front and back covers are reversed images of each other. Like looking through a mirror or a finger etching on a frosty pane of glass.

Do these images exist? Are they objects, or are they concepts? See the article "What Is a Shadow?" to find out!

The Rational Scientific Method

By
MonkEmind

Hypothesis, Theory, and Conclusion:
A Rational Scientific Method of Inquiry

In science, a definition is a limitation or restriction on the use of a word. Scientific definitions are rational, non-contradictory, unambiguous terms that are consistently used and narrowly defined by the person who is making the hypothesis. We use adjectives to modify nouns (objects) and adverbs to qualify verbs (concepts). Science, in general, and physics in particular are about the physical, those things which have physical presence; what is real; things that exist.

To exist means to have shape and location, that is, an object with a location; something, somewhere. We visualize objects and we explain concepts. We do not explain objects - we point to them, or illustrate them. We explain phenomena. The scientific method is based on hypothesis and theory. The conclusion is left to each individual. The hypothesis includes the statement of facts, the definitions of key terms, and the objects. The hypothesis describes the phenomena and illustrates the objects, defines the key terms, then makes the assumption(s). Assumptions are statements of the facts - not the facts themselves. Assumptions are neither true nor false. One does not define objects; one illustrates them. The theory explains the phenomena of the hypothesis. Everyone must decide for themselves.

Each individual concludes that the theory is either possible or not possible. Science is about explaining. Science in general and physics in particular are about physically present objects.

Understanding the difference between objects and concepts allows one to make a rational conclusion about the key terms and the statement of facts at the hypothesis stage of the scientific method.

Proof is for math. Science never proves. Science is about physical reality. Math describes abstract dynamic concepts, whereas science illustrates static physical objects, and explains phenomena. Math is NOT the language of science, illustration is.

A hypothesis stands on its own. It does not matter who agrees. The hypothesis should illustrate the objects, define the key terms, and present the statement of the facts; the assumptions. The theory would then explain the phenomena of the hypothesis. There is no correct or incorrect hypothesis - it is an assumption. It is either rational or not. If it is rational, we accept the assumption(s) of the hypothesis. Predictions and observations are opinions and are extra-scientific.

Hypotheses are assumptions and theories explain the hypotheses. We form a conclusion that the theory is either possible or not possible. This is why, in science, it is crucial to understand the difference between objects and concepts, nouns and verbs, adjectives and adverbs, hypotheses and theories. We can say: I see a field of corn. The corn stalks wave in the wind. I have a dust particle in my eye. BUT fields, waves and mass-less particles are concepts in math which do not exist in physical reality therefore should not be presented in the hypothesis. We describe or illustrate objects in the hypothesis.

We explain concepts in the theory. We never explain objects, we describe them, illustrate them, or point to them. Sometimes we eat them!

"Insofar as mathematics is exact, it does not apply to reality; and insofar as mathematics applies to reality, it is not exact." – Einstein Sometimes Einy made sense!

The mathematical physicist uses ambiguous or contradictory terms inconsistently. He or she confuses objects with concepts, nouns with verbs, adverbs with adjectives, and hypotheses with theories. Reality does not depend on human perception or observation. It is because the human senses are limited and flawed that science must be objective. The scientific method should be observer independent as much as possible. A rational key term never invokes an observer. Although our senses are limited, there is no limit to our intellect. One must apply rationality, reasoning and critical thought at the conceptual stage in the hypothesis. Precision is precious. Defining key terms is critically important. Understanding the difference between concepts and objects is essential in dealing with science. We make this clear with our definitions. In science, one must be able to visualize the concrete object. Objects must be illustrated in the hypothesis. The objects are the actors, the key terms make clear the meaning of the script, and the statement of facts sets the initial scene for the theory. The dynamic concepts in the theory are describing the phenomena of the hypothesis. The hypothesis is a photo (static), the theory is a movie (dynamic).

Each person takes away their own conclusion as to whether or not the story was possible. Most important are the key terms and these have meaning as defined by the theorist. In science, one can only use objects that can be illustrated in the hypothesis. If it cannot be illustrated or visualized, then it is not real and has no physical presence. What is not physical has no place in science.

Science, especially physics, is conceptual. Technology, which is mostly trial & error, is empirical.

Planes that fly, microwaves that heat and GPS devices that measure your position work primarily through trial and error because of technology, not because the theories that they are founded upon are "correct."

The problem lies in the confusion between objects and concepts. There is no good way to discuss General or Special Relativity, Quantum Mechanics, or String Theory until point, line and plane can be defined and understood. Math attempts to describe dynamic concepts by moving numbers. Physics is about reality. What exists, physically present objects with location are made up of matter. These are static and can be photographed or illustrated. But we must be able to define what "exist" means.

Universe: matter (atoms) and space (nothing)

Concept: the relationship between two or more objects

Object: that which has shape

Space: that which does not have shape

Exist: matter + location

Location: the set of static distances to all other objects

Motion: object + 2 or more locations
Theoretical physics, Newtonian physics, ToR and QM don't explain anything, they describe. These theories "predict" or describe, but do not explain. It is not interesting that Newton tells me an apple falls at 9.8 meters per ft per second per second. I want to know why. I can point at an apple and say, "Look it is falling fast." So what? What is the physical medium that attracts objects to each other? That is the question for science. Math "predicts" how fast something falls to the ground, but it says nothing about why it falls.

"Since the mathematicians have invaded the theory of relativity, I do not understand it myself anymore."—Albert Einstein

Ptolemy "predicted" to a high degree of accuracy the position of the planets in the solar system, but he had the earth in the center. That does not help explain why the planets orbit in elliptical paths and don't fly out into space.

What about these "predictions?" If I observe an apple fall a few times and measure the speed and distance traveled, I can "predict" how fast an apple falls. What does that tell me? It does not tell me when an apple is going to fall. Now THAT would be a real prediction. Something that already happened, a consummated event, is described and should then be explained.

Something that we have observed happen repeatedly can lead us to think that there is a high degree of probability that it will happen that way again. But that is not really a prediction - it's an educated guess.

Belief, truth, evidence and proof are not part of the scientific method; it is observer-independent. Experiments and observation are extra-scientific. Science, especially theoretical physics, is conceptual. Technology, mostly trial & error, is empirical. Here's the root of the problem with the currently taught scientific method: It all revolves around simple misunderstandings of basic physical reality brought on by the inability to determine the difference between an object and a concept, and the inability to precisely and consistently define terms upon which a theory depends.

At the root of the Relativity and Quantum Mechanics problem is Euclidean geometry. Because the point, the line and the plane are not defined, or, are defined ambiguously using abstract concepts instead of objects. They do not represent actual physical reality! That is a rather shaky basis on which to form the physical "laws" of the universe.

Rational Scientific Method
Hypothesis: Illustration of objects; definition of key terms; statement of the facts, the assumptions.

We assume in the hypothesis stage. If the assumptions are rational, then we can proceed to the theory. The objects of the hypothesis are described or illustrated; a photograph-static.

Theory: explains the hypothesis; phenomena such as motion or process.

Conclusion: possible or not possible? Everyone decides for themselves.

If the key terms of the hypothesis are ambiguous, circular, synonymous or contradictory, then the theorist should throw out the hypothesis, or present precise, rational definitions of key terms upon which the hypothesis depends.

The theory is where we present a "movie" or series of illustrations of the phenomena, or process involved in explaining the hypothesis. Then, and only then can we form our conclusion.

If we conclude the theory is irrational, and therefore not possible, we throw the theory out.

If we conclude that the theory is possible, then we may publish a paper. If we conclude that the theory is possible, but does not provide the complete explanation, we form another hypothesis based upon the theory and build upon it. The flat earth becomes the round earth, which becomes the oblate spheroid... Once the theory is presented, science is done!

The conclusion is left up to each individual: Possible or NOT possible!

What Is Rational?
By MonkEmind
and Fatfist

What is a rational definition?

What is a rational explanation?

What is a rational Person?

A Rational definition is unambiguous, non-synonymous, non-circular, non-contradictory and can be used consistently throughout a presentation.

What is a rational explanation? "An explanation which:

"1) has no contradictions

"2) doesn't reify objects into concepts

"3) does not perform verbs on concepts

"4) does not use concepts to perform verbs

"5) uses unambiguous and consistent terms, thus being grammatically correct and understandable

"6) can be visualized, illustrated, and can be put on the big screen as a movie without any missing frames. If it cannot be visualized, then it cannot be understood because it contradicts reality.

"A rational explanation is unbiased, observer-independent and understandable.

"This is pretty standard stuff, which unfortunately, eluded philosophers and logicians during the Dark Ages and even to this very day!" - Fatfist

What is a rational Person?

A rational person is willing and able to follow an explanation to the only conclusion to be had---possible, or not possible.

When one understands that there is only possible or not possible, it takes the desire out of holding onto a particular idea as the "ultimate truth" or position to have. A rational person understands that better explanations may come along at any time and is not invested emotionally in the outcome or emotionally connected to any particular conclusion.

A religious, or dogmatic person, on the other hand, depends on prefabbed conclusions which are built upon premises that are etched in stone. The result is dogmatism, stubbornness to change, and refusal to accept any contradictory position to their own, rational or not." - MonkEmind

What Is Rational and What Is Logical?
by
Fatfist

RATIONAL:
The word "rational" applies specifically to statements, which are the linguistic outputs of our thoughts. We say that a statement is RATIONAL, when it meets ALL of the following criteria:

1. It does not reify concepts into objects and does not attempt to apply motion to concepts or to nothing.

2. It can be visualized, illustrated and can be put as a movie on the big screen without any missing frames.

3. Every crucial term that is referenced can be defined unambiguously and used consistently.

So clearly, what is RATIONAL has, without question, nothing to do with truth, absolutes, proof or logic.

Rational thought is 100% without question, completely divorced from LOGICAL thought. They are in completely different categories.

1. Logical thought (truth, proof) is founded upon artificial axiomatic rules conceived by humans. Logical systems come in 2 forms which are represented in either SYMBOLIC or CONTEXTUAL languages. They are only DESCRIPTIVE, and never prescriptive.

They cannot prescribe or explain reality. This means that as a language, logic (truth, proof, etc.) deals exclusively with concepts, and not objects. Hence, logic can NEVER have any EXPLANATORY power. Logic can never be used to EXPLAIN why natural phenomena happen a certain way because it deals with concepts (verbs & adverbs), and NOT objects (the nouns of reality)!

Logic is only descriptive.....it can only DESCRIBE the subjective and opinionated observations a human made via his physically limited sensory system. This is why logic is 100% DIVORCED from reality!

2. Rational thought is predicated upon reality, Mother Nature's realm, which can only be critically reasoned by humans.

It is not based on artificial axiomatic rules or any systems of logic.

It is based on the realization that the universe consists of either: SOMETHING (that which has shape), or NOTHING (that which doesn't). There is NO conceivable middle ground between shape and no-shape, or any other option.....ever! This is reality, and this is what the word RATIONAL is predicated upon!

What is rational is not subject to someone's opinion. This is the objective criterion demanded by reality....Mother Nature's realm. Rationality is necessarily predicated upon OBJECTIVITY, without the injection of the opinions of human observers. Otherwise, it will degrade to an issue of ordinary speech, religion, opinion or the Symbolic Logical Religious sect known as Mathematical Physics, which is 100% dependent upon the subjectively biased observations, opinions & descriptions of humans. Not a single rational explanation can ever be offered by any discipline which is based on LOGIC.

If somebody claims that their God Theory is RATIONAL, then they had better know what they're talking about. If they cannot visualize their own theory, and make a movie out of it, then they have NO clue of what they are talking about. They do not have a "rational explanation" by any stretch of the imagination.

For example, a God Theory would have to show in frame #1 of the movie, God with his magic wand, existing in nothing; i.e. no space, no matter, no void. In frame #2 and later frames, the movie would have to show how God creates space. We cannot just have space magically appear in frame #2. How did God do it?

Similarly, in say frame #100, the movie would have to show how God created the first bit of matter.

It is irrational to just have matter appear in frame #100, when it didn't even exist in frame #99. The movie is supposed to explain all these issues by showing how no-shape can surreptitiously acquire Length, Width, and Height, and turn into shape/form with internal structure.

An irrational explanation consists of a scene that cannot be visualized or imagined. An irrational explanation is one that the proponent cannot illustrate or convert into a movie.

Find MonkEmind books on Amazon.com. In Paperback and E book format; for Kindle, Nook or PC, as well as in audiobook formats; audible by Amazon and iTunes

The Rational Scientific Method

By
Bill Gaede

Introduction

Science is the body of papers accumulated over the years that follow the "scientific method." The "scientific method" is a rational way of presenting explanations. This rational way of presenting explanations consists of three steps: 1) hypothesis, 2) theory, 3) conclusions. Each step is based upon the one prior. If an explanation has been given in accordance with the scientific method, then it is a scientific explanation and a rational explanation – thereby an explanation which one has good reason to believe. Finally, we will then conclude by explaining what the scientific method is not.

Step 1: Hypothesis
A hypothesis is comprised of a) exhibits, b) definitions, and c) a statement of the facts/assumptions.

a) Exhibits

- Exhibits are objects and only objects.
- This is because you "need to be able to visualize" what is being referred to later in Step 2: Theory.
- Exhibits are evidence.

b) Definitions

- Define the key terms which make or break the theory.
- Definitions are limitations placed on a word's utility or extent.

This is because by defining as many differentiating qualities about a concept as possible one may reduce the breadth of the word being used and zero-in on the description with the most necessary precision. A perfect definition is one that has been refined to the point where everyone interprets exactly the same thing. Only define concepts not objects.

- Instead of defining objects, point to the image and name it.
- Object is a word which is a category including only those words which represent shapes. The objects should be exhibits.

c) Statement of the Facts/Assumptions

- Describes an object or tells us what happened in an event.
- Addresses all necessary how questions, and does not address why questions.

In order to understand conceptually how this is done, the following digression regarding fact hood must be made:

- *Fact/Truth = The Universal Movie*

Visualise a movie consisting of movie-frames/photographs of the entire universe: Every frame contains every single object in existence.

- Each object is distinct from one another, with definite location.

There is no real movement in such a movie, and it would be perceived only due to memory of past frames when the movie is played. Therefore, each frame contains only shapes.

The universal movie is an endless collection of frames with shapes arranged inside.

> Such a movie would be fact hood; An uninterrupted sequence of locations of every atom in the universe.

> Statement of the facts consists of the selection of the clips from the Universal Movie/Fact necessary to keep in mind for the theory in question to be understood.

> Clear frames (A) in are objective evidence / exhibits / evidence, and filled in frames (B) are subjective testimony from the observer concerning the event or object in question. The frames can be judged to be objective only in as much as they correspond perfectly with regards to the locations of objects in fact, and not with judgements.

> A= Universal Movie / Fact / Truth
> B= Statement of the Facts/Assumptions and Evidence: Subjective testimony (filled in frames), and objective declaration / Exhibits / evidence (clear frames).

The Statement of the Facts is not truth, or fact, but opinion which may be "relatively objective" ("true") or subjective (false). However, ultimately, the Statement of the Facts is always subjective.

The Statement of the Facts is an opinion of what the facts are, presented in the form of an assumption, but is not fact itself.

They are your opinion about what you saw, not what you thought you saw or wanted to have seen.

An exhibit, is fact or evidence, when pointed at and named, it is a statement of the facts.

Fact is neither theory nor knowledge, and is not established democratically.

There is no observer of fact. It cannot be observed. Observation implies subjectivity, so in trying to understand fact, "kill the observer," i.e. do not invoke observer-dependent concepts.

Rates, ratios and relations may be components of the statement of the facts.

You must be able to take the statement of the facts at face value, and if they do not appear logically invalid or inconsistent then they may be rejected as grounding for your theory.

Step 2: Theory

- A particular version of how or why events happened.

- Speculation about some of the missing frames of the Universal Movie.

- Possible clips of the Universal Movie inferred from assumptions and reasoning.

- Must be visualisable.

- Must follow from the assumptions and evidence, i.e. is logically valid.

Theories fill in the blanks in order to infer how or why events in the assumptions occurred. They do not become fact if accepted by the majority. If accepted by the majority, theories become assumptions in the statement of the facts, i.e. they undergo a Kuhnian "paradigm shift" and are taken for granted.

Stage 3: Conclusions

- The verdict and opinion on the theory. Conclusions are only objectively: possible or <u>not</u> possible.

- Synthesises inferences

- Tells us what experimentation or data we may need to verify the theory

- Multiple opinions may be formed and debated over given that both parties forming these opinions accept the theory.

Example of the Entire Scientific Method

Hypothesis (assumptions)

Exhibits: Earth, Sun, space

Definitions: object, motion, space, mass, planet

Statement of the facts: The Earth goes around the Sun.

Theory (explanation):

The reason that our planet orbits the Sun is that the Sun's mass warps the space around it. The Earth is a little ball rolling around an enormous roulette.

Conclusions (opinions):

1st opinion: The Earth touches space and we should be able to run an experiment to verify warped space.

2nd opinion: Warped space is a mathematical abstraction and is beyond experimentation.

What the Scientific Method is not

1) *The purpose of the Scientific Method is not coming about with a description, but an explanation.*
Contrary to the historic development of science, a proper theory which consists of mathematics is not a theory at all because mathematics only has the power to describe – and then only quantitatively. With the RSM, we hypothesize objects to explain phenomena.

2) *The Scientific Method, after the hypothesis, is observer free*
That is to say, the theory is not an observation or description, but an explanation. The theory is not acceptable just because you can observe it to be true, in which case it is self evident, the theory is acceptable because it offers a rational explanation. This is the definition of objective.

3) *The purpose of the Scientific Method is not for facilitating prediction.* A prediction is a description about a consummated event in the future.

We make predictions when we think we know, based upon experience, that an event will occur.

The fact that we can explain why an event will occur inherently facilitates prediction in the future, but prediction is not the purpose of the scientific method itself, the purpose of the scientific method itself is associated with the past, namely the mechanisms and causes (i.e. objects) which may be invoked as explanations for an event.

4) *The purpose of the Scientific Method is not for facilitating experimentation or tests.*
The results of experimentation or tests only have a place in the hypothesis as evidence, they themselves may never be the theory itself, or "proof" that your theory is correct. If evidence proves a theory, then there is no theory.

5) *The purpose of the Scientific Method is not to give us knowledge of the truth.*
The purpose is to explain and fill in the gaps in the hypothesis stage. Whether you accept it as truth or not is your opinion, a matter of debate, and confined to the conclusions stage. The same goes for being a "proof" that your beliefs are true. If the truth or proof is so clear as a result of your engagement in the Scientific Method, then there would be no need for theories in the first place, but pure hypotheses which can be accepted at face value.

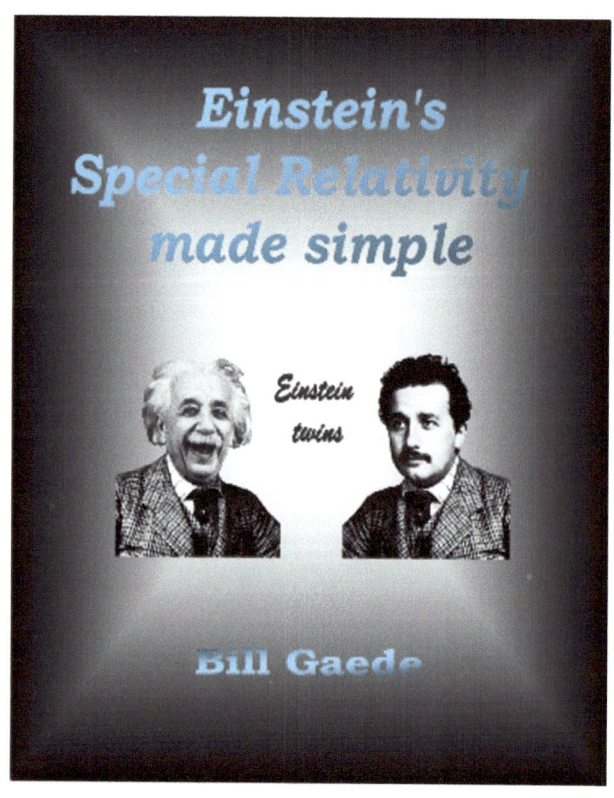

Rational Physics
By
MonkEmind

- Objects have shape
- To exist is to have shape and location
- Space can not become matter and matter can not become space
- Matter and motion are eternal
- All phenomena are the result of surface to surface contact between objects

All words fall into one of two categories.

THERE ARE OBJECTS AND THERE ARE CONCEPTS. It is very important in any conversation to understand the difference. In science it is crucial that all terms, which make or break an argument or presentation, are defined scientifically In the hypothesis.

Put in terms of physics, there are physical objects and phenomena. Physical objects, that is, objects which exist, have shape and location. Physics is the study of physical objects; something somewhere, objects with location. A circle is an object because it has shape. It can be illustrated. It does NOT have location with respect to all other existing objects. Draw a circle on a piece of paper with a pen. What exists is the ink and the paper. The circle is an abstract concept geometers invented. The shape is defined by that which allows us to conceptualize what is inside and outside of a border; space. The one criteria common to all objects is shape. This is an either/or situation. Objects have shape, space does not. Therefore, space is not an object!

Cut a circle out of a piece of paper and see that we still have a "shape," but it is now a ring of paper. The ring has the mutually orthogonal directions of length width and height; the three dimensions of reality. The ring exists because it has location with respect to all other existing objects.

When the paper ring was cut out of the piece of paper there was a hole, or space "left" in the piece of paper where it was removed. Did we destroy paper? Did we create space from paper? Put the ring of paper back. Did we create paper from space? No, of course not, we simply changed the shape. Matter can not be created or destroyed. It is eternal! By matter I mean the sum of all objects being comprised of atoms.

Object: that which has shape

Space: that which lacks shape

Without the concept "space" we would not be able to conceptualize any existing objects because there would be no border or separation between them. We can take our ruler and measure the hole in the paper. We have not measured space; we have measured distance from one place on the paper to another place on the paper.

ALSO, IF THERE WAS NO SPACE, OR SEPARATION BETWEEN OBJECTS, MOTION WOULD BE IMPOSSIBLE. If we accept the definition for object, we see that all objects are finite. If there was no limit, or border to an object, no other objects would be possible, and therefore there could be no motion. What is motion? It is two or more locations of an object.

It is apparent that there is matter and motion. This is the default position one must take. If anyone says there was a beginning to matter (creation) or motion (first cause) then the onus is on them to show how this can be. Naturally, we understand that matter and motion are eternal. For how can nothing, that which lacks shape, or space, become some "thing." In other words, how can zero dimensions instantly become the three dimensions of reality? As we understand from our example above about paper, it is impossible for space to become matter and matter to become space. Also, how can there be a beginning to motion? If it is necessary for a first cause of movement, what moved the first mover? We go on forever recursively and never get to a first cause!

Lastly, we understand that all phenomena are the interactions between objects. We drew a circle on paper and we cut out the circle with a pair of scissors. The objects were pen, paper, and scissors. The phenomena were draw and cut. There could be no draw without paper and pen. There could be no cut without paper and scissors. There can be no phenomena without two or more objects interacting. The interaction involves contact between surfaces. Can you name an exception?

Science & Technology - Conceptual & Empirical
By
MonkEmind

This author considers science as purely conceptual.

Evidence, proof and facts are not considered scientific. They are not part of the Rational Scientific Method of inquiry because they are dependent upon the limited sensory system of man and are therefore considered opinion. We reserve our opinions for after the hypothesis and theory is presented and it is called the conclusion. The conclusion is ours alone.

What is technology? We often see the phrase, "Science & Technology." Technology is empirical, evidence based and mostly trial and error. The difference is often debated. A general consensus might be that science and technology are interdependent but separate endeavors.

From Ask.com we find the following: "Science is a way of practicing knowledge, as well as the knowledge itself, whereas technology is the application of science, particularly to industrial or commercial objectives. Technology can also be defined as the scientific methods and materials used to achieve industrial objectives."

Before we continue, let's look at some common definitions.

According to Dictionary.reference.com, science is "a branch of knowledge or study dealing with a body of facts or truths systematically arranged and showing the operation of general laws: the mathematical sciences. 2. systematic knowledge of the physical or material world gained through observation and experimentation."

Knowledge, facts and truth are the opinions of man, and as such, have no place in science or technology. That is part of religion and philosophy. Science explains and technology builds.

One would think that if science and technology are interdependent yet separate that there would be a friendly relationship and a clear division of labor between the two. Yet this is not the case. Scientists, engineers, social scientists, philosophers, historians, policy makers and the public all have self interests. Because of this government and other agencies like NASA have Science and Technology departments just to help solve related issues and coordinate resources and development across the disciplinary boundaries.

The rivalry is one issue and the focus has been mostly on that. Just take a look at all the scholarly articles devoted to it. As an electrical engineer, and later a software and then hardware engineer, it was obvious to me who did the work and who got the credit.

Wiki Answers tell us about the relationship between science and technology:

"Science discovers FUNDAMENTAL INFORMATION about how the universe works. Technology is the practical application of that information, or knowledge. A computer is an example of technology; in order to invent one, it is necessary to know a lot of fundamental science. Science sets the stage for technology, which produces useful devices. There would be no laptops without the FUNDAMENTAL DISCOVERIES of science."

We hear all the time that science is responsible for this and that. I've been told that if not for Quantum Mechanics the transistor would not have been invented, if not for relativity, GPS would not have been possible and so forth. This is not the case, as we shall see.

Many great discoveries had nothing to do with the "science" behind it.

"Chance favors the prepared mind." - Louis Pasteur

Are we to believe that serendipity is part of the scientific method? We are told that it is:

"Serendipity means a 'happy accident' or 'pleasant surprise'; a fortunate mistake. Specifically, the accident of finding something good or useful while not specifically searching for it. Indeed, the scientific method, and the scientists themselves, can be prepared in many other ways to harness luck and make discoveries." - WIKI

Charles Goodyear accidentally spilled a mixture of rubber, sulfur and lead onto a hot stove creating vulcanized rubber.

The engineer, Wilson Greatbatch, used the wrong value resistor and the circuit he was working on pulsed like a heart beat giving him the idea for the pace maker.

Alfred Nobel accidentally discovered dynamite when he dropped Nitroglycerin in sawdust. The sawdust soaked it up, stabilizing it and making it useful as dynamite.

Alexander Fleming discovered Penicillin a day after failing to clean up his work station.

Benedictus dropped a flask that had contained a liquid plastic that had evaporated. The flask didn't shatter and safety glass was the result.

While trying to make artificial quinine, Perkins made the first synthetic dye superior to any of the natural dyes available at the time. Chemistry quickly became a money making enterprise. Later a German bacteriologist by the name of Paul Ehrlich, used the dyes in a different manner. He used them in immunology and chemotherapy.

Potato chips were an accident, as were popsicles, ice cream cones, and Coca Cola. Smart dust (silicon chips) used for sensors was an accident as well.

Saccharin was discovered because a chemist didn't wash his hands and chemical got on his wife's dinner rolls.

Percy Spencer was using a vacuum tube and aimed it at various items in the lab and accidentally melted a chocolate bar in his pocket. The beginning of the first microwave oven!

Wilhelm Roentgen was tinkering with a device when he noticed a fluorescent light flickering. He started putting various objects in front of it and discovered he could see the bones of his hands. The X-ray machine had its beginning! Later, when experimenting with X-rays, Becquerel accidentally exposed a photographic plate with a uranium rock and then with the help of the Curies, discovered radioactivity.

Chemist Leo Hendrik Baekeland, trying to make a shellac alternative, produced a material in one of his experiments called Bakelite. He was going to use it to make phonograph records but found it could be used for many other things. Plastic is derived from it today.

Lysergic acid was absorbed through the skin of Albert Hofmann and he got a buzz. Timothy Leary was happy about that because LSD came about because of it!

Phizer discovered that a drug they were using in a clinical trial for heart conditions, although useless for that purpose, was great for erectile dysfunction. Viagra was born!

Smallpox vaccination, clinical use of insulin for diabetes, and the Pap smear...all a result of serendipity. As were post it notes, Cellophane, and Velcro. So were Play-Doh, Stainless Steel, the Ink Jet Printer, and Vaseline ...all a result of serendipity.

"The seeds of great discoveries are constantly floating around us, but they only take root in minds well prepared to receive them." - Joseph Henry

Thank God for open minds and thank God for Viagra and Vaseline!

While these are all good examples of accidents, luck, and happenstance meeting up with open minded, observant men and women, it points to something very important. It points to the difference between science and technology, conceptualizing and observing.

The Free Dictionary tells us that conceptual is relating to mental conception and gives a use as "conceptual discussions that antedated development of the new product."

Experiments: Are They Part of the Scientific Method?

By
MonkEMind

No, Kiera, they're not! Experimentation is observation based and therefore extra-scientific.

A lot of money and resources will be saved when this is understood by the scientific community and a corresponding cooperative division of labor is established between science and technology.

Aristotle conceived of a spherical earth long before Eratosthenes "predicted" it with math, and Magellan "proved" it by circumnavigating the globe.

Let's take a look at experimentation, and why it can go terribly wrong.

So what is being taught as the scientific method? This is it in a nutshell:

Research the question

Form a hypothesis

Conduct an experiment

Analyze the data

Draw conclusion

Communicate results

http://www.slideshare.net/mrmularella/developing-a-hypothesis-and-title-for-your-experiment

"Research the question" means see how other people have answered it. "Form a hypothesis" means repeat what you have read or heard about the question. "Conduct an experiment" means design a method that describes the hypothesis. "Analyze the data" means use observation to confirm your hypothesis. "Draw conclusion" means decide

The same source defines empirical as: "Relying on or derived from observation or experiment: empirical results that supported the hypothesis." The second definition is revealing: "Guided by practical experience and not theory, especially in medicine."

Does science depend on accidents, happenstance, or serendipity?

Does empiricism depend on theory? No!

Rational Science takes this a step further. Empiricism is extra-scientific. That is, it is NOT part of the Rational Scientific Method of inquiry.

Why?

Because the human sensory system is limited but the ability to conceive is not.

It is technology with its empiricism, experimentation, and its trial and error that is responsible for our so-called scientific advancements. Science played a very minor role.

The mathematical theorist, for instance, may believe that their calculations confirm, and GPS proves, Relativity. They therefore take credit for our GPS system. The calculus only confirms that as the satellite orbits the earth caesium atoms are being stressed differently than the caesium atoms in the clocks at the ground stations. The atoms were stressed. Time was not warped as proposed by the ridiculous theory of relativity!

The transistor was discovered in Bell Labs by engineers tinkering with components and not because of some theory or theories relating to the ridiculous notions of Quantum Mechanics!

The purpose of science is to explain and that depends on conceptualization. The usefulness of technology is in designing and building. This is accomplished by experimentation, testing, and trial and error which depend on observation.

Clearly, science and technology are two different things altogether.

The problem with modern scientific method is that they confuse the two. Scientists today confuse nouns with verbs, concepts with objects, hypothesis with theory, and science with technology.

whether your hypothesis is right or wrong. "Communicate results" means to tell others what your conclusion is.

This from Science Projects.com

"The Job of the Scientist is to study the surrounding world and explain why the world is the way that it is."

Good so far! Science should be about explaining.

"The way that this is carried out is by experimentation. The methods for producing experiments comprise what is called THE SCIENTIFIC METHOD."

http://www.science-projects.com/SciMeth.htm

Hmmm. Not so good. Observation is what brought us to this point. We saw something interesting and we want to understand the phenomena. It's circular to use observation to "explain" observation. Not only that, but our senses are flawed and limited (as discussed in the Rational Scientific Method). Also, when we look closer we discover that experiments aren't explaining anything anyways, they are describing.

"When preparing to do research, a scientist must form a hypothesis, which is an educated guess about a particular problem or idea, and then work to support it and prove that it is correct, or refute it and prove that it is wrong.

"Whether the scientist is right or wrong is not as important as whether he or she sets up an experiment that can be repeated by other scientists, who expect to reach the same conclusion."

http://answers.yahoo.com/question/index?qid=20060905151511AAkOd9L

We are told that the idea is to understand cause and effect using a controlled experiment which utilizes controls and variables. But understanding does NOT come from guessing, experimentation, right, wrong or proof. Understanding comes when one can conceptualize the objects and rationally explain the phenomena. Experiments describe. Descriptions don't explain anything. We describe objects, we explain phenomena. We point at an apple, describe it, or offer a photograph of it. We attempt to explain WHY the apple fell onto Newton's head. We don't describe it falling at so many meters per second squared.

A kindergarten child understands that if the apple falls from the tree it moves real fast and lands on the ground. BUT WHY? Why doesn't the apple fall up into the sky? This is what everyone really wants to understand.

Who cares how many persons can repeat an experiment if we don't actually understand anything.

If you research "experiments gone wrong" you will generally find the type of accidental discoveries discussed in the article "Science and Technology." These are the good results of bad science.

If you have been around the interwebs for any time, you have likely run across the many charts showing the correlation between the number of pirates and global temperature. Of course this is done in fun to illustrate the point that "correlation does not imply causation."

FAILED EXPERIMENTS

General R.G. Dyrenforth, the concussionist, represents another such example and brings it closer to home for us Texas boy… er MonkEs. Although not a General, or a Commissioner of Patents, as he claimed, he was able to convince many people into believing one could blow up explosives in the air and cause it to rain.

Others had believed that rain followed artillery. Plutarch, Napoleon and Edward Powers erroneously believed this as well. In 1871 Powers wrote a book about it entitled "War and the Weather." He even convinced the US government into paying him $2,000 to make it rain. Dyrenforth was tasked with the job.

It's a funny part of Texas history, where experiments conducted in the heat of a Texas summer went comically wrong. Trainloads of dynamite and gunpowder were sent to Midland.

The "General" told New York Times reporters that it "is a matter of cold fact, based on my experiments, I know that rain can be produced."

Newspapers such as the Chicago Tribune, New York Sun, and the Washington Post wrote articles about man's will to control nature for his own purposes. Most of these reporters didn't even go to Texas. However, Texas Farm and Ranch, and Farm Implement News reporters were there to witness the experiment. Canons shot explosives into the air, and kites and balloons carried various explosives up in to the wild blue yonder to be blown away by strong Texas winds and scattered around exploding at the wrong time and lighting all kinds of things on fire.

Scientific American magazine later said the experiments were "an expensive farce."

Of course, Powers never considered the probability of rain for any given location when there wasn't a battle, and the fraudulent General only succeeded when Mother Nature decided SHE was going to provide rain... in spite of his silly antics. People apparently had enough of him after he blew out some windows in a San Antonio Hotel and wiped out a nearby mesquite tree.

Chernobyl was a failed emergency shutdown experiment.

In 1962, Tusko the elephant was given LSD (3000 times the human dose) just to see what would happen. They just guessed this amount even though an elephant is about 90 times the weight of a human. The elephant died. The 'scientists' told Science magazine, "It appears that the elephant is highly sensitive to the effects of LSD."

Look up the Monster Study where children were made to stutter just to test the idea of positive and negative reinforcement.

Check out the experiment by Psychologist Winthrop Kellogg and his wife. They raised their newborn son David along side a chimp named Guo until they saw that their son was more like a chimp then the chimp was like a human.

There are thousands of examples, but perhaps the most famous is the one by B.F. Skinner who raised his baby in a box. This one, however, went exactly as expected. Still according to the public, when they found out, it was terribly wrong. Keep those psychologists away from me please!

Here's an experiment gone way wrong. A Florida teenager is charged with a felony because of her "failed" experiment. She was in a science class mixing various household chemicals when the Eight oz. water bottle exploded (no one was injured). Kiera Wilmont was taken away in handcuffs and expelled from her school.

What really went wrong? Was it the curious girl's experiment? Was it the fact that she was unsupervised in the science lab? I submit that nothing really went wrong based on what is being taught in these science labs. The problem is when science classes teach the scientific method involves experimentation.

Why are these failed "scientific experiments?" Because experiments are not part of the scientific method! They are part of technology's trial and error.

Is this science? When Thomas Edison was interviewed by a young reporter who boldly asked Mr. Edison if he felt like a failure and if he thought he should just give up by now perplexed, Edison replied, "Young man, why would I feel like a failure? And why would I ever give up? I now know definitively 9,000 ways that an electric light bulb will not work. Success is almost in my grasp." And shortly after that, and over 10,000 attempts, Edison invented the light bulb. – Yahoo Answers

No, this was clearly trial and error. Eventually a fantastic result, but not scientific! Today scientists are still unable to explain electricity or light.

When persons believe science is based upon observation and experimentation, and true and false, this is the sort of thing that they may end up believing.

The Real Story Behind America's UFO Connection and Area 51

"There is a story involving the U.S. military's Area 51 in Nevada that is so incredible it makes seasoned psychics and remote viewers turn ale when they look at it"

Not if one dispensed with ridiculous notions of psychics and remote viewing which has yet to be rationally explained. Scientific method based on observation, government secrets and UFO reports abounding, is it any wonder the gullible public falls for this sort of nonsense?

"Aliens are real. They exist in a parallel universe that some people call the astral plane, or sub-space. It exists all around us. The aliens are mostly beings of light to us. The best way to get a good look at them is to learn how to leave our bodies and enter the parallel universe where they live."

Parallel universes are part of science fiction, not science. People would understand this if they weren't indoctrinated with the ridiculous SM being taught today as a result of superstitious magical foo held over from ancient times.

"We humans are so powerful we can kill the aliens with a mere thought."

Wow! And how do we "know" this?

Evidence and experiments. Yep, that's good enough for me!

"There is evidence in ancient rock carvings, ancient writings and aboriginal legends that the aliens made contact with humans in the distant past. ... That we show different skin colors and appearances also suggests that the experimentation either was done by more than one visiting alien race…"

Evidence and experiments. Yep, that's good enough for me!

"But there is something even more important. It seems that we exist for a while in solid bodies that living creatures in sub-space do not have. And while we exist in them, our bodies give us the ability to touch, taste, smell, hear and fully enjoy the things around us."

Exist is still not defined scientifically. It's no wonder we hear this sort of thing all the time (From: The Mind of Amos Donahue).

Psychic powers, parallel universes, alien and faster than light spacecraft, light beings that can inhabit our bodies, these are all things that persons believe based on evidence, true, false, observation and experimentation.

Once one understands the Rational Scientific Method, they will never fall for this sort of thing again. Why? Because what is possible or not possible is not dependent upon observation, experimentation or belief of man!

If you are thinking well, yeah, of course those things are silly. That is only because you don't have the evidence before you that YOU need, unlike these others. Obviously, they have all the evidence that THEY need. Evidence, facts, true, false and belief are all observation based...just like experimentation.

Science has its Many Worlds interpretation, where science fiction has its parallel universes.

Science has its mass-less particles, where science fiction has its ghosts.

Science has light particles that arrive before they leave, where science fiction has its time travel.

What's the difference between science and science fiction? It's hard to tell, if you don't understand the Rational Scientific Method.

Word Magic V1.1
From Grinning Monkey Publishing

Tired of being pushed into the corner with scientifically accurate words and phrases? Well, no more, astound, befuddle and bamboozle your opponents with a working knowledge of the top misleading words in the human lexicon. With words like energy, field, wave and life you'll never be pinned down again!

With WordMagic V1.1, you can say anything, and it is guaranteed to be true, correct and factual!

You can be the life of the party! You can be unequaled in irrelevancy, fabulously fallacious, and deliciously discrepant. Learn how to impress your family, friends and neighbors with counterfactual vernacular and meaningless jargon.

Whether at the seminar or symposium, conference or confabulation, stun your co-workers, colleagues and collaborators, with inconcise, indeterminate and unspecific Magical words of misrepresentation and miscommunication.

With WordMagic V1.1 you'll be able to define ambiguously, circularly, synonymously, and inconsistently, thousands of authoritative, popular words from our tested and true Word Magic Lexicon.

But Wait! There's more! Order now, and we'll include, for no additional charge, our very specially narrated CD version. Listen to the smooth, sonorous voices of folks like, Stinking Hawking and Dick Dawk as they teach you how to dodge and weave through any presentation, debate, conversation or convocation.

And we're still not done! For the next 100 persons who order WordMagic V1.1, we'll include, free of charge, the additional Musical CD, "WordPlay." Relax during an office power nap, or send yourself off to Slumberland listening to these favorite hits: Double Talk, Ambiguity, Dubiety, Incertitude, Tergiversation, Equivication, Polysemy, Double-entendre and Enigma.

WordMagic It's not just for scientists anymore!

Do you suspect that something is wrong with what you read about in the so-called physics journals and popularization magazines? Don't black holes sound a bit magical? Doesn't Big Bang come across as Creationist? Does the establishment's "explanation" that there are many copies of you, each in a different universe, make you wonder about the state of mind of the theorists? If so, perhaps it is time for you to consider an alternative.

Why God Doesn't Exist (WGDE) presents a fresh perspective on the nature of light, gravity, magnetism, the atom, and the workings of the Universe in general that is rational. It is rational because each theory is illustrated. You can understand the mechanism by just watching the videos that accompany the book. It is also mainstream theories, you will not be asked to believe in the movement of concepts (e.g., transfer energy, move "a" mass, carry "an" interaction). WGDE is for intelligent laymen who simply want to understand the causes and mechanisms that underlie physical phenomena.

To obtain a paperback, Paypal to bill@youstupidrelativist.com
USA/Canada: US$ 30   Europe: 30 Euros

Bill Gaede Interview
by
MonkEmind

MonkEMind, author, editor and publisher caught up with Bill Gaede for this one of a kind interview. In spite of his busy schedule, Mr. Gaede, Father of Rational Science and the Rope Hypothesis, afforded us this unique opportunity, allowing a glimpse into the mind of this century's most interesting and diverse scientist, laying the foundation for this and the next three issues of The Rational Scientist magazine.

MonkE: Who are you? Where were you born and what's your background?

Bill: My name is Bill Gaede. I was born in Argentina, and I worked primarily as an engineer in the semiconductor industry.

MonkE: Wikipedia says you were an industrial spy; that you were caught and served time in prison for stealing proprietary materials from INTEL and AMD where you worked as an engineer. After you served your time you were deported from the United States. Why should we trust anything that you say?

Bill: Let's deal with these two issues separately. First, the criminal case:

1. I pled guilty to stealing. This is different than saying that I stole. The correct word is "copying." I *copied* written material and distributed it. But copyright is handled under tort law. In fact, the US Attorneys told the press that the reason they agreed to a short sentence is that there were no laws to prosecute cases like mine. They simply winged it and stretched the interpretation of existing federal law to incorporate my actions.

In order to trigger a FEDERAL criminal case of stealing, the stolen material has to be US Government property. That wasn't the case here. No Government property was involved.

In fact, the property in question did not even belong to the private companies that complained to the US Attorney.

To this day the *manufacturing processes* of integrated circuits (chips) are not publicly registered or disclosed. Companies such as Intel and AMD would be at a loss to prove in a court of law that "their" manufacturing processes have been "stolen."

The only existing records are internal specifications. The companies can at best fire an employee and sue him for violating a *civil* agreement of confidentiality which the employee must agree to and sign when hired.

I pled guilty to "stealing" because:

a. The prosecutors promised to send letters to the Immigration and Naturalization Service (INS) (which they did) asking them not to deport me (which the INS did anyway).

The legal basis was that the federal prosecutors considered this a case of "moral turpitude" (non-deportable) rather than a deportable case under "aggravated felony."

The US Attorney's Office handles both criminal and civil (deportation) cases. The US Attorney's Office handling the immigration case didn't follow the "recommendations" of the US Attorneys that dealt with the federal criminal case. They took the position that my actions constituted an aggravated felony.

What is more interesting and more fundamental is that both the US Attorneys in the federal criminal case and the INS knew that I was an illegal alien.

The FBI report provided to me under Rule 16 clearly states on the first page that I entered the US as a tourist and overstayed my visa for about 25 years!

b. They went after my wife. The plea agreement states clearly that:

GOVERNMENT'S PROMISES

9. The United States Attorney's Office for the Northern District of California agrees:

(a) ...

Further, no charges will be filed against defendant nor his wife for any activities or things possessed in the District of Arizona relating to this investigation, such as materials related to Intel and Advanced Micro Devices, as such as various forms of identification documents.

Are Mainstream Science Magazines Really Science?

THE RATIONAL SCIENTIST

APRIL 2018

NO! It's NOT science, Mildred!
In this issue you'll Discover why.
It might be popular but it's NOT Science!
It might be American, but it's not Scientific.

Energy, Waves, Mass, Fields: These are Concepts not Objects. Learn the difference and never be fooled again!

Troubling Issues in a Silicon Valley Spy Case

By CALVIN SIMS

BUENOS AIRES, July 7 — On a brisk May afternoon last year, a frantic Argentine man carrying a stack of papers and videotapes knocked on the doors of several American newspaper offices here.

"I'm a spy, and I think they're going to kill me, so I want you to know what has happened," were the first words out of his mouth.

The man identified himself as Bill Gaede, an engineer who claimed to have stolen secrets about computer-chip technology from companies in the United States and passed them to the Governments of Cuba, China, Iran and others. And now, he insisted, he was being tailed by the Central Intelligence Agency and by Argentine intelligence agents.

So remarkable was Mr. Gaede's tale of political and industrial espionage and of his eventual conversion to being an F.B.I. informant that few people believed him. Not even when Mr. Gaede, whose first name is Guillermo, though he goes by Bill, provided audio and videotapes of encounters that seemed to support his story.

As it turns out, many of Mr. Gaede's contentions have in the last year been shown to be true. Mr. Gaede (pronounced GAY-dee), who is 42, recently began serving a 33-month prison sentence, now at the Santa Clara County jail in San Jose, Calif., after pleading guilty to Federal charges of mail fraud and interstate transportation of stolen information. Mr. Gaede says that a plea bargain he struck allows him to remain in the United States after completing his sentence and prevents prosecution of his wife, who he says assisted him in his activities.

His booty, the Government says, was trade secrets from the Intel Corporation,

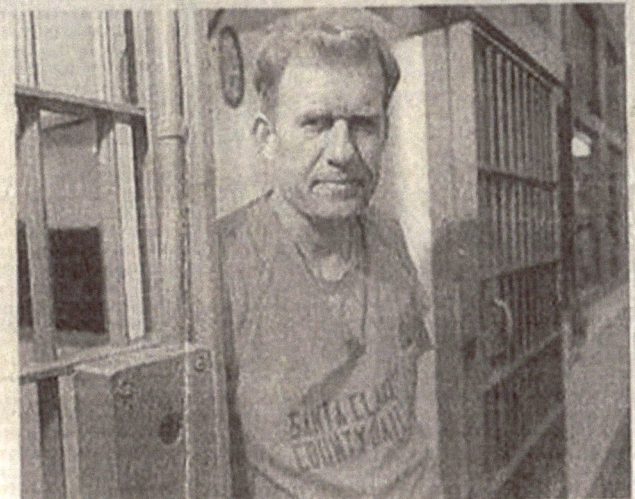

Guillermo (Bill) Gaede is serving a 33-month prison sentence after striking a plea bargain on Federal charges that he stole chip-making plans from the Intel Corporation. He says he sold the information to several foreign countries.

the largest maker of semiconductors in the world. The Government said those secrets were worth $10 million to $20 million.

And now, whether one believes Mr. Gaede's contention that the F.B.I. condoned his espionage or accepts the view that Mr. Gaede acted alone and simply managed to slip through Intel's security net, his account of foreign spying might give pause to any chief executive in Silicon Valley. "There are so many foreigners working in these companies that this sort of theft is bound to happen again," Mr. Gaede said in a telephone interview last week from the jail.

The F.B.I., which last year declined to comment on the case, now acknowledges having had regular meetings with Mr. Gaede while he was stealing trade secrets

Continued on Page C2

Case of High-Technology Industrial Espionage Raises Troubli

Continued From First Business Page

from Intel and his previous Silicon Valley employer, the chip maker Advanced Micro Devices Inc. But the agency denies Mr. Gaede's contentions that he was an informant and that the F.B.I. sanctioned his espionage as a means of infiltrating Communist governments.

George Grotz, a spokesman for the F.B.I.'s office in San Francisco, said in a recent telephone interview that the bureau warned Intel in June 1993 that Mr. Gaede had a history of espionage. Intel says it never received that warning.

Mr. Grotz said that during the F.B.I.'s long relationship with Mr. Gaede, the bureau considered prosecuting him for criminal activities on at least three occasions. But the Government did not bring charges, he said, because there was no applicable Federal economic espionage statute.

Still, "it was our responsibility to notify Intel of a situation we thought was important," Mr. Grotz said of the warning that the F.B.I. said it issued in 1993.

Mr. Gaede is highly critical of the bureau.

"I admit that I broke the law and that I should serve time in jail," Mr. Gaede said. But "the F.B.I. was an equal participant in the crime because its agents instructed me to continue stealing information so they could catch the Cubans."

"The F.B.I. is not clean in this case, and neither are the companies that I stole from," he said, because their security was inadequate.

The crime to which Mr. Gaede pleaded guilty did not involve a transfer of Intel secrets to a foreign government but, instead, what Intel executives say was an attempt to sell the information to his former employer, Advanced Micro, in May 1994.

Mr. Gaede contends that he was framed by F.B.I. agents, who he said sent the material to Advanced Micro to build a legal case against him. But he does acknowledge stealing the information and says he sold it to China and Iran, through various contacts, for "large sums of money."

Many troubling questions remain. How, for example, did Mr. Gaede, an Argentine with forged United States documents, gain access to the trade secrets of the country's largest computer chip makers? Do United States companies have adequate security measures to prevent future acts of espionage? Did F.B.I. agents fail to properly monitor Mr. Gaede, who continued to steal trade secrets even after he had briefed agents on his past thefts?

Shades of 007 in a scheme involving Intel's 486 and Pentium chips.

Mr. Gaede came to the United States in 1977 on a tourist visa and stayed on by using forged documents. Mr. Gaede, who said he was a communist ideologue, began working for Advanced Micro as an engineer in 1979. He said he began stealing the company's plans for its computer chips in 1982, giving the stolen information to Cuban officials in hopes that it would help the Castro Government develop its own semiconductor industry.

The US Attorneys in Arizona had started an investigation of my wife, attempting to involve her in my case and, thus, the Government significantly raised the stakes. By including her, my "stealing" case could quite easily be converted into a RICO case (Racketeer Influenced and Corrupt Organizations > 1 person conspiracy) which typically carries a sentence of 20 years for both of us. In exchange, they offered me 3 years (of which I had already served 1) and no deportation. It was an offer I couldn't refuse.

It's the way the US Government does justice when it really wants you.

As to whether people can trust anything I say, this has to do with what I say and not with who I am or what I did. A mechanic may be a rapist or a murderer. However, if he can explain exactly how a car works (and fix yours), in what way do his crimes invalidate his explanation?

MonkE: What are your credentials? You were an engineer in the computer industry. What makes you think that you are qualified to propose alternative hypotheses for light, gravity, electricity and magnetism?

Bill: If you get into a fight with an unknown fellow on the street, you don't ask him for his credentials as a boxer before you struggle with him. The winner is decided objectively. It's the man who remains standing.

It could be argued that a person who is college-educated is more qualified to theorize than the common man. And yet we have Rocky Balboa training in a meat shop and giving the world champ a run for his money. It could also be argued that a mathematician has the skills and training to do sophisticated calculations that give him a significant edge over a layman. However, these skills have nothing to do with his ability to theorize about the nature of our Universe.

We do not need Math to make an assumption about the structural nature of light and the atom. Albert Einstein has as much authority as the bum living under the bridge to tell the world what an atom looks like in reality. The fact that Einstein studied Math does not give him any superpowers in this respect. Likewise, the fact that a geneticist knows his DNA back and forth and can tell you how much of it matches that of Neanderthals gives him no authority to tell you that Neanderthal slept with your great grandmother thousands of years ago. People tend to confuse lab and computer experts with theorists. These activities require different skills.

Indeed, a college-educated individual is someone who had to be able to repeat what the mainstream published in books and that his teacher makes him memorize in order to pass the course. Being able to spit out facts learned by rote is a little different than reasoning something out for yourself.

MonkE: Haven't there been hundreds of so-called Grand Unified Theories before? What makes yours so special?

Bill: The attempt to unify "forces" – specifically push and pull – has incongruously drifted into Mathematics. As a result, Quantum Mechanics has ended up with four forces that the mathematicians call: electromagnetic, gravity, weak, and strong. Yet gravity and the strong force are pull forces and the electromagnetic and the weak forces are push forces. There is no form of contact that we can imagine other than push and pull. And we certainly cannot use equations to elucidate the invisible mediators of these forces. There is only one way and one way only to unveil the physical architecture of an invisible entity and that is to make it visible. The proponent has to tell the crowd what he proposes to do gravity and light with.

My proposal is one that anyone can visualize. It doesn't mean that people have to agree or believe in it. It just means that they can understand the theory that is founded upon it. They can visualize the mechanism because I can make a movie of the entity moving and they can just watch the film. Straight forward!

What is a mathematician going to put on the screen for energy or for a 0D 'particle' or for four-dimensional warped spacetime? To this day there is not a single mathematician on Earth that can illustrate a hydrogen atom for you despite that Quantum has been around for hundred years. Now, how difficult can it be to draw a particle next to another: an electron bead next to a proton bowling ball?

MonkE: What is rational science?

Bill: Rational Science differs from what the mathematicians do in that it is done with objects. An object is that which has shape. We illustrate these shapes on a frame and make a film strip. This is a theory: a movie of the mechanism. The invisible objects such as the mediator of light and gravity, the mediator of magnetism, the atom itself, etc., are made visible in order to make a movie of the mechanism so that the spectators understand what the theorist is proposing. It doesn't mean that they will believe or accept or agree with what is being proposed. It just means that they will understand the mechanism.

In contrast, the mathematicians who purport to do "science" and call themselves "scientists" harp on proof and truth and fact. They claim such powers because the worldly wisdom is that Math is infallible. People extrapolate that if the Math is rigorous and the mathematicians check the equations and find no fault in them, then the "theory is proven." What has been "proven" is a mathematical DESCRIPTION: an equation of sorts. This is entirely divorced from the qualitative EXPLANATION of the mechanism. Physics has to do with explanations and not with descriptions. Mathematicians are not physicists.

MonkE: What's to be gained by going back to the old way of doing things when the current empirical method of observation, experimentation and duplication has worked so well and has produced all our modern technology like computers, cell phones and GPS which makes humans so unique among Earth's creatures?

Bill: There is a great misconception in the general public as well as among professionals that science has something to do with technology. Therefore, in order to answer this question rationally, we must begin by defining these two terms…

Science: rational explanations

Technology: development of gadgets through the process of trial and error

Technology is the realm of engineers, not of scientists. An engineer doesn't have to understand or explain how or why his device works. If it works, he's done! A missionary can show the natives in the jungles how a magnet picks up pins. That's technology. He doesn't have to tell them how or why it does so. In fact, not a single so-called "physicist" on Planet Earth can today.

The misconception has spread that computers wouldn't work if Quantum Mechanics turned out to be wrong or that GPS wouldn't work if General Relativity were wrong. These are gross misconceptions because not a single computer engineer can tell the crowd *what* a magnetic field *is* or how it does its magic.

And GPS was developed by trial and error through experiments such as Gravity Probe A and the Hafele-Keating Experiment. These experiments do not tell you what caused the difference in time. They just confirm that there was a difference in time. The clocks in the GPS system are merely adjusted in accordance with the results of the experiments.

MonkE: Over 400 years of scientific research has already determined that light has the properties of both a particle and a wave. Clearly, science has proven the particle and wave nature of the atom and of its constituent parts, including the proton, electron and neutron.

Bill: Yes. Whatever light IS, this enigmatic mediator BEHAVES at times in a vibrating or undulatory manner and at others like an impacting stone. Of course, we cannot simulate these behaviors with waves or with particles. So what the mathematicians did was merge the wave and the particle into an unimaginable entity they call "wave-packet." This so-called wave-packet is not supposed to be taken literally. There is no physical object called wave-packet. The term wave-packet refers, instead, to the behaviors of light rather than to what the mediator of light looks like. In contrast, a rope is an imaginable entity. A rope does simulate both wave and particle behaviors. You torque a rope and anyone watching from the side sees waves propagating along the rope much like when you watch a barber's pole twirl. The rope presses on whatever it is connected to at its ends. We have wave and particle behaviors in one mechanism.

MonkE: I understand that you have some different ideas about why species go extinct. Can you briefly outline your proposed extinction mechanisms?

Bill: There are two mechanisms that I propose for how species disappeared in the past: one for background extinctions where a single species disappears all on its own and another for mass extinctions where a family or order of species disappear all of a sudden in geological terms. A background extinction occurs when the population pyramid of a species overturns and the last generations lose their genetic diversity. A mass extinction occurs when the population pyramid of a class of plants overturns. There are fewer and fewer varieties and their numbers also dwindle to near extinction levels. They are replaced by newer, more efficient and advanced plants. The herbivores that forged a relationship with the ancient regime find fewer and fewer niches to squat. Their varieties and numbers fall as well until the entire food chain disappears.

MonkE: Why do you think this applies to man? Man has the ability to predict the future and adjust accordingly.

Bill: Man has a level of intelligence that no other animal has been able to emulate. This gives Man the ability to predict the future, make adjustments through technology if necessary, and solve problems. However, there are things that humans cannot do despite their intelligence and technology because intelligence and technology are incongruous tools to use for such problems. In what way can either intelligence or technology be used to overcome death of an individual or stop humans from aging or incite them to have more children or prevent them from losing genetic diversity? In what way can intelligence and technology cure us of the need to eat or to breathe? Technology is not a panacea as most casual theorists seem to think. Humans cannot do anything they can imagine or solve all problems, certainly not with technology.

MonkE: Environmentalists, economists and transhumanists disagree with you. What do you have to say about their proposals?

Bill: Transhumanists are people who have over-indulged in technology and allowed their minds to take them beyond the realm of reality. The modern world has overwhelmed their feeble brains. They talk about the *singularity*: a point in future "history" after which technology and knowledge will begin to expand exponentially. The rationale behind this is that artificial intelligence (AI) will help us discover more technology and concepts and solve more and more problems. The robots and androids we manufacture at the factory will be so sophisticated that they will have the ability to tackle issues that we petty humans cannot even imagine.

I can only suggest that these pitiful individuals turn off their TV sets. They have watched too many Twilight Zone and Star Trek episodes. Transhumanists no longer have their feet on the ground.

The reality is that humans are about to become extinct. They will become extinct because our global economy is going to crash. By *economic collapse* I don't mean what economists and environmentalists mean. I specifically mean that "money will be no more"! It is money which no longer will have value.

A CURIOUS
CHAIN OF EVENTS

24

No one on Earth has proposed this before, and in fact all of these economists and environmentalists argue for hours against me that it is impossible for money to disappear.

Therefore, anyone who claims that I'm no different than climate change theorists such as Guy McPherson has no idea what my theory is about. When Guy McPherson mentions the term "economic collapse," under no circumstances is he referring to the disappearance of money. Neither he nor anyone else ever mentioned that money will be no more. Whenever an environmentalist or conservationist mentions "economic collapse" he instantly begins talking about resources and overpopulation and peak oil and reusable energy. I am talking about something else. I am talking about the artificial economy that Man has created. I am saying that at some point there will be a stock market crash coupled with high unemployment coupled with indebted governments and unlimited money creation. These trends and events lead to the crashing of money itself. WHEN that happens, it's all over. It's over because agricultural companies have at that point no pecuniary incentive to plant or harvest. Transportation companies have no pecuniary incentive to deliver food to the cities. The *vital* system that maintains us alive is irrevocably broken. Governments don't even have the leverage to enforce laws or induce soldiers to put order. Quite the contrary! Soldiers and lawmen will be using their handguns to kill and steal when all hell breaks loose. This mechanism has nothing to do with what climate change lobbyists are proposing

Therefore, transhumanists, economists, and environmentalists have lost sight of what actually keeps us alive: food. We all take food for granted. Most of these individuals have in the back of their minds that technology solves everything. When they talk about economic collapse, they merely think that there will be some kind of recession or depression like in the past. And since we've always come out of these difficult economic moments, we will do so the next time a depression arrives. These people think of "economic collapse" as a 'normal' business cycle from which we always recover. Therefore, chicken little does not warn you that it's the end of the global economy. He trains you on how to make the most of the post-depression recovery. He asks you to invest in gold or bitcoin because you'll be rich *when* the economy recovers.

Knowledge and Prediction

by MonkEmind

"The meaning of knowledge is known to all, and I think it means acquaintance with a fact, science or technique for example." - Andy

Fact: opinion from authority. Authority: Self appointed expert on a particular subject

"I see what you are insinuating. Thanks." - Andy

Typically we see something like this from Oxford dictionary:

"knowl•edge - facts, information, and skills acquired by a person through experience or education; the theoretical or practical understanding of a subject."

If one can rationally explain it, then they understand it. However, much of what self-proclaimed authorities and scientists claim as knowledge is nonsense. How can anyone possibly understand the irrational (impossible)?

Mainstream physicists use "predict" and "know" irrationally. When they say "equations predict," such as, $F=ma$ predicts how fast an apple falls from the tree, they are actually describing consummated events.

When they say they know the apple will fall from the tree at 9.8 ms ^{2} and then hit the ground, they mean they "predict" based upon past experience. If a raven swoops down and grabs the apple before it hits the earth, so much for their knowledge, and so much for their prediction.

Our "knowledge" is based in large part on observation, and observation is subjective, being built from limited human senses.

Because of this and because of "an ongoing collision between belief and reality" science necessarily must remove the observer from the scientific method. Knowledge, experience, and belief are opinions, and have no place in science or physics which studies what is physically real.

Knowledge has nothing to do with reality. Knowledge is a hallmark of religion. Knowledge just means that you have made up your mind about something or other. There is only one way for you to show that you know something: you run an experiment and prove your knowledge to yourself.

- know/knowledge: The ability to predict the result of an experiment without error. Knowledge only has to do with the future.

- explain/explanation: To state the causes of a phenomenon (consummated event). Explanation only has to do with the past.

Science is the study of existence/reality using the Scientific Method. There is no provision for knowledge in the scientific method.

The scientific method only deals with consummated events which are rationally explained. Astrology deals with future events which are predicted and allegedly known. In science we strive to understand. We do not confuse understanding with knowledge.

What Is A Shadow?
By
MonkEmind

What is a shadow? A hologram? A reflection?

Are they objects? Do they exist? How about the ring shape in the pool of water below a water fountain?

Surely there were reflections and shadows before any living thing existed on earth. Surely holograms will continue to exist long after the last hairless ape is gone. So these are more than just concepts, aren't they?

Quick review of definitions:

Object: that which has shape

Concept: relation between two or more objects (or nested concepts)

Exist: object with location

Imagine a row of birds sitting on the ground and the shadows that they cast. Imagine a mountain reflecting in a lake. Imagine the scene in Star Wars where R2D2 is projecting the holographic image of Princess Laya. Imagine a cascading fountain forming a ring in the pool below. Imagine the edge of a forest and the shadow it casts onto the ground. NOW, Keep these images in mind throughout the following discussion:

Surely birds cast shadows before we imagined or saw them, before there were cameras to photograph them, before there were people to make cameras. And if you could zoom into a single eye of one of these birds you would see what we call a reflection of the other birds, and the shadows in the pupil of that bird.

Is Princess Laya an object? Does she exist? Is she a concept? What about a reflection of a shadow? Is that an object?

How is the reflection of the mountain like the shadow of the birds, the ring under the fountain and the holographic projection of Princess Laya?

According to Wiki: "Reflection is the change in direction of a wavefront at an interface between two different media so that the wavefront returns into the medium from which it originated.

"A shadow is a space where light from a light source is blocked by an opaque object. It occupies all of the three-dimensional volume behind an object with light in front of it. The cross section of a shadow is a two-dimensional silhouette, or reverse projection of the object blocking the light.

"Typically, a hologram is a photographic recording of a light field, rather than of an image formed by a lens, and it is used to display a fully three-dimensional image of the holographed subject, which is seen without the aid of special glasses or other intermediate optics. The hologram itself is not an image and it is usually unintelligible when viewed under diffuse ambient light."

So what is alike in all of these is that it takes two or more objects to create. Is it light? The ripple or waves required the water drops and the basin of water. Can there be ripples under a fountain without any light to see it? Yes. Then light is not what is common to all of these. What is?

There is interaction between two or more objects, as in all phenomena. A shadow, ripples, holograms, and reflections are not stand alone objects.

Without discussion about light specifically, we note there are objects involved in all of the phenomena: "Reflection is the change in direction of a wavefront at an interface between two different media so that the wavefront returns into the medium from which it originated."

"a hologram is a photographic recording of a light field"

"A shadow is a space where light from a light source is blocked by an opaque object."
All of these involve two or more objects. BUT there are serious flaws in all of this because there is no clear distinction for what is an object and what is a concept.

So far we have dealt mostly with objecthood. The next crucial question is, "Does the ring, the bird's shadow, the mountain reflection and Princess Laya exist?" AND the answer is NO! The fountain and the water exist. The sun, the bird and the ground exist. The mountain and the lake exist. The air molecules and the projector exist.
The skeptic will now say that if the shadow or reflection, the ring and the holograph are there with or without an ape to conceive of it, how can it be a concept? BUT LET'S NOT EQUATE CONCEPTS AND PHENOMENA EXACTLY.
A phenomenon is the result of surface to surface contact between two or more objects.

A concept is the relation between two or more objects.

In the human language all words resolve to either concepts or objects.

In reality all phenomena resolve to surface to surface contact between two or more objects.

Man discovered certain phenomena (now they are considered concepts) such as space, length, distance, shadows, reflections, etc.
The mountain and the lake, the sun, the bird and the ground, the water drops, water and fountain; projector and air molecules, are that which have shape. Therefore, they are the objects. The reflection, the shadow, the ripples and Princess Laya are the phenomena. They are NOT objects, nor do they exist!

The shadow appears to our eyes because the bird's body has shifted the frequency of sunlight down to a lower frequency than that which is not a shadow. Atoms are pumping and torsion occurs along their interconnecting ropes whether we are there to perceive it or not. More on this later.

Always keep everything in CONTEXT: A single tree casts a shadow. BUT a forest can cast a shadow. A tree is an object for purposes of stating the squirrel climbed the tree. A forest is an object for stating the plane crashed into the forest.

We can say loosely that the shadow has shape for purposes of saying "the bird cast a shadow that looked very much like the bird," but really it is the ground that is the object we are seeing (Keep in mind that we are also seeing but less aware of the molecules in the air that are shaded between the bird and the ground).

This is akin to saying the circle we drew with crayon on a piece of paper has shape. And so we might refer to the circle as an object because it has shape, but it is the paper and the crayola that have shape. We are again being loose with our words here because "a circle" can not possibly exist by definition. We also understand that the crayola and paper exist because it is they that have shape and location with respect to all other existing objects. Further, we understand that "circle" is a mathematical construct by definition, an abstraction that is impossible to find in nature or produce.

There is a clear distinction between objecthood and existence that needs to be kept in mind at all times. We need to be very precise with our use of terms in science, but only nearly so with casual conversation. The point is, we need to put everything in context. We need to be clear on the use of our terms in a scientific discussion and avoid ambiguity.

At best we can say shadow, like a circle, is an abstract object, because shadow does NOT have shape, therefore is NOT an object, and does not exist in a scientific context.

Now, let's be very specific on what a shadow IS.

Light is a torsion signal which occurs along em ropes which extend between all atoms in the sun, the bird and the ground. Pumping, vibrating atoms of the sun induce torsion which occurs along interconnecting ropes at varying frequencies.

Frequencies are the number of links per unit length of rope. When the bird appears to block the light shading the ground, it is actually changing the frequency along ropes interconnecting the atoms in the sun to the ground.

Our eyes are also connected to atoms in the sun, the bird and the ground. Therefore, we perceive this difference in frequency as it is relayed to our eyes and translated in our visual cortex.

What a dog sees (by this I mean torsion signals relayed via the eyes), or a bird sees, or a bee sees, is the same as what we see, but likely is perceived (what is relayed to the brain and what the brain does with it) differently. Atoms in a dog's eyes are connected to atoms in the sun, the bird and the ground as well but they only have yellow and blue sensitive cells in their retinas and so they won't perceive exactly what we do as hue and tint (colors blended with black and colors blended with white) is different.

Our visible light spectrum is about 390-700 nanometers, but a bee's is from 300-600 nm. So their visible spectrum includes blue-green, blue, violet, and ultraviolet. Birds not only see everything we do, they have additional color cones in their retinas and can also see UV. If we could see everything along the em spectrum (all frequencies) then what we would probably perceive is white. We would not be able to distinguish one object from another by color. BUT whether we can perceive something or not has no bearing on the reality of an object. We have to assume the invisible physical entities which are responsible for phenomena we wish to explain.

When we understand that all phenomena is the result of surface to surface contact between two or more objects we understand that shadow, ripple, hologram and reflection are phenomena.

Psychophysicists consider shadows and reflections as objects that exist for purposes of shape reconstruction, hopefully, with the goal of being able to better render 3D computer images. Circles are drawn by geometers and considered as objects for purposes of understanding shape and space.

But physics is concerned only with objects that exist. That which has shape and location with respect to all other objects; that which is physically present. So for purposes of a hypothesis one illustrates their objects. We note that they have shape, so that they may be considered objects. Space lacks shape so can in no way be considered an object. Therefore it can not be used in any hypothesis as an object.

Conclusion:

We may assume that Rutherford's planetary model of an atom is an object for purposes of a hypothesis (he drew it after all), but if it has no explanatory power we throw out any theory that is based on that atomic model.

A circle can only be said to be an abstract object when we see that it has been defined:

"a round plane figure whose boundary (the circumference) consists of points equidistant from a fixed point (the center)."

Clearly, circle is a concept (some"thing" called an abstract object).

In the case of shape reconstruction we understand that the reflections and shadows are not really objects, but representations of objects that psychophysicists wish to study. In the case of the circle we understand that we have an "abstract object" which can not possible exist at all in nature or even be reproduced. The shadow is not the object and the circle drawn on paper is not the abstract object. AND neither are really objects at all! In the case of the former, it is a phenomenon, in the case of the later it is a concept.

Whereas all phenomena are concepts, not all concepts are phenomena. There are higher order concepts, or nested concepts like marriage, for example. This union is not an object and neither are "husband" and "wife." Marriage is the relationship between a husband and wife (or in the case of same gendered couples; wife/wife and husband/husband). Phenomena is the surface to surface contact between two or more objects.

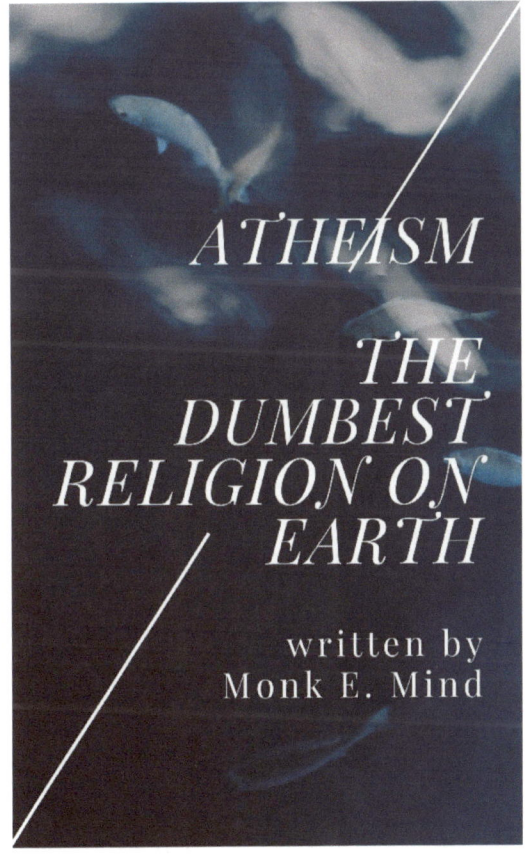

The Rational Scientific Method

by
David Robison

Several months ago I wrote an article on Bill Gaede's Thread Theory. Here, I would like to provide a formal justification for the "visualization" approach to physics described in that article. I will do that by outlining an alternative version of the Scientific Method called the Rational Scientific Method. The number of proponents of this method has been slowly but steadily growing over the years. I don't think there is 100% agreement in every last nuanced aspect of the method, but overall there is a common understanding. Below will be my understanding of it.

From now on I will refer to the Rational Scientific Method as the RSM, and the current version of the Scientific Method as the CSM.

RATIONALITY

If one is going to have a "rational" method they need to define the term "rational." Dictionary.com defines it as such:

1.   agreeable to reason; reasonable; sensible

2.   having or exercising reason, sound judgment, or good sense

3.   being in or characterized by full possession of one's reason; sane; lucid

4.   endowed with the faculty of reason

So, rational = reasonable. As I'll argue later, this illustrates why dictionaries have no place in science. They are not designed to give scientific definitions, but to document common usages for everyday speech. Almost every word in a dictionary has multiple definitions, and oftentimes they are synonymous or circular.

In this case, the definitions are synonymous, they simply point to another word without narrowing the usage. In addition, what's reasonable to one person may be unreasonable to another. Opinion cannot be the basis of a scientific term.

So what is the objective criterion? The only objective criterion for a definition is that it must be able to be used consistently.

If a theorist uses a key term at point A in his presentation, he should use it to mean the exact same thing at point B and C. Otherwise, he's being inconsistent and leaving key words open to interpretation by the audience. People will come away with different understandings. This is the foundation of Religion — the ability to utilize concepts like belief, truth, proof, evidence, knowledge and fact with respect to theories, where key terms have been left open to interpretation.

Rationality, on the other hand, is about clarity and consistency. The antithesis of rationality is ambiguous, vague, unintelligible, incoherent, inconceivable and unimaginable.

Rational – describes any communication in which the key terms have been defined to the point where there is only a single possible interpretation for them (i.e., where everyone understands the same thing).

A definition is a limitation on the usage of a term. A scientific definition is one which limits the usage of a term down to a single possible interpretation. Rational communication then is when the key terms have been provided with scientific definitions.

The key terms are those which are essential to communicating the meaning of the theory. No one expects you to provide scientific definitions for "a" and "the." If the key terms were defined differently, it would completely alter the meaning of the proposal and thus it is crucial that they be clearly delineated. The key terms are those which make or break your proposal.

EXPLANATION VS. PREDICTION

One of the major differences between the RSM and the CSM is that the RSM is strictly a formal means of proposing explanations. There is no provision for predictions whatsoever. The domain of the RSM is past events.

The CSM, on the other, hand holds prediction as the hallmark of a good "theory." Simultaneously, however, they talk about explanations. The two become muddled together. The result is that the key terms of the CSM itself are left open to interpretation.

Take the term hypothesis.

"A hypothesis is a proposition that is consistent with known data, but has been neither verified nor shown to be false." – Wolfram Alpha

"An idea that proposes a tentative explanation about a phenomenon or a narrow set of phenomena observed in the natural world. The two primary features of a scientific hypothesis are falsifiability and testability…" – Encyclopedia Britannica

"A scientific hypothesis is the initial building block in the scientific method.

Many describe it as an "educated guess," based on prior knowledge and observation.

"While this is true, the definition can be expanded. A hypothesis also includes an explanation of why the guess may be correct." – Live Science.com

The definitions show no clear distinction between explanation and prediction. A hypothesis allegedly incorporates both simultaneously. Yet, the two are radically different. In the context of physics, the failure to draw a clear distinction has led to explanations being considered "proven" or "factual" because they are associated with an equation which can make accurate predictions. This has happened despite the explanations being irrational.

The RSM, on the other hand, clearly distinguishes explanation from prediction, and only deals with the former. In order understand the RSM approach, it is necessary to first define the critical term "object."

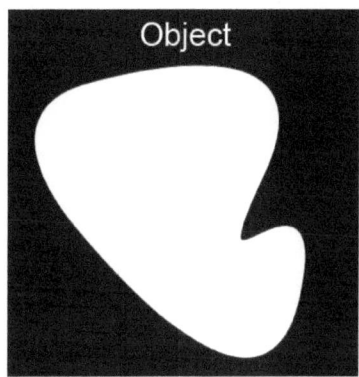

An object is separated by its outline from its immediate surroundings. Whether something has shape, or not, is a black and white matter. It either does or does not. There is no conceivable in between possibility. There is no such thing as "half shape" or "partial shape." An object does not in any way blend with its surrounding. The boundary is definite and sharp.

Shape is the only inherent quality of an object and is common to all of them. An object can have shape regardless of whether a living entity is there to observe, sense or perceive it in any way. This means an object's shape is innate and observer independent.

Object – that with shape

SPACE

Even a lone object in the Universe, all by itself, must still have a background in order for it to have shape. Space is the unbounded environment which is necessary for the lone object to have a contour.

Space – shapelessness (i.e. vacuum, emptiness, nothingness, formlessness, the void)

Pursuant to the above notion of the term "rational", the objectively distinguishing characteristic of these definitions is that they can be used consistently throughout a scientific dissertation and can drive out the possibility of miscommunication to the greatest extent humanly possible.

Ultimately, visualization lies at the heart of all conceptualization. Objects can be visualized and this act precedes any defining of concepts. Upon inspection, the mind outlines any relations and assigns labels to them. With objects as the basis, one can begin to critically reason ever more complex and nested concepts and delineate them with scientific definitions.

THE INTELLECT AND THE SENSORY SYSTEM

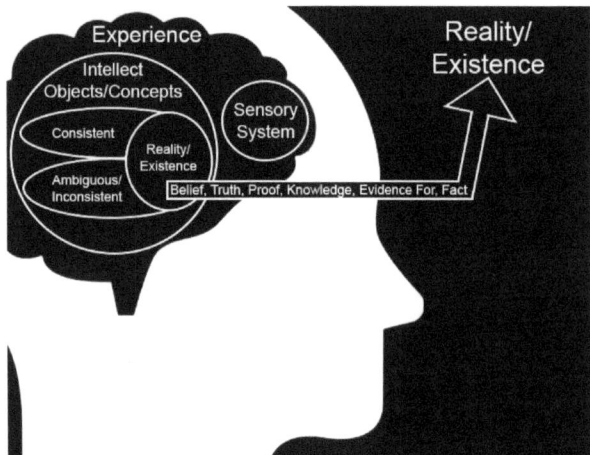

The above diagram is not meant to be scientific, but is just a helpful tool to illustrate the thinking process at work here.

From our individual perspectives, we are completely trapped within the world generated by our neural activity. All we ever have to draw upon is our immediate experience. That experience can be broadly divided into two distinct faculties: the intellect and the sensory system. The intellect is responsible for conceiving of objects and understanding concepts. The sensory system is responsible for processing evidence via the senses.

The intellect is distinct from the sensory system because one can always conceive of multiple possible theories to explain any given body of evidence (observations, measurements, experiments, equations). In that way, the intellect is not contingent upon the sensory system. The intellect can draw from the unlimited possibilities, whereas the sensory system is a direct line to the sensory organs. Not only can the intellect conceive of different explanations for the same evidence, but it can conceive of the existence of objects which could never be directly gleaned through the senses — like that of atoms or of galaxies beyond the visible universe. Whereas the sensory system is limited by the constraints inherent in using physical organs to interact with nature, the intellect does not have these limitations.

No matter how big or small, or distant, the intellect can conceive of the shape of the objects. The sensory system, however, doesn't say anything about what is. It deals only in appearances and is always open to broad interpretation by the intellect.

ASSUMPTIONS

The fact that we are trapped within our own experience means that we have no choice but to conceive of ideas about the external world beyond our immediate experience.

The sheer act of formulating ideas about external reality assumes that there is something forever beyond what is immediately available to us.

Thus, despite having access to evidence through the senses, we will always be ultimately ignorant about the external. Because of this we have no choice but to make assumptions about existence.

Assumption – a statement taken at face value for the purpose of understanding a theory

Making an assumption does not mean that the statement is believed or regarded as true. It means that the statement is being temporarily entertained for the sole purpose of establishing understanding.

Understanding always precedes any consideration of belief or truth. A claim has to be intelligible before one can even begin to decide whether they think it's true or false. Therefore, the assumptions must be asserted and analyzed prior to any assessment of belief or truth.

Assumptions can be categorized into two distinct camps: those in which the key terms have been defined consistently, and those in which they have not. This is not a matter of belief, truth, proof, evidence, knowledge, fact or any other subjective concept which relies on the sensory system. The individual is immediately aware of the assumptions via the intellect and can critically reason whether or not the definitions for the key terms can be used consistently.

In the depiction above I have divided the intellect into consistent and ambiguous or inconsistent. Within both are existential assumptions. There is no way to avoid making assumptions about existence in the course of proposing physical explanations. The only choice one has is between clarity, and consistency, or ambiguity and inconsistency.

The devil's advocate may argue here that the two options are not distinct, that it's more of a blurred mush. Inevitably, the theorist will fall somewhere in between clarity and vagueness. This is not the case, however. Shape is the quality which stands out against the muddle as being clearly distinguished. An object lies in stark contrast to its immediate background. They are of clear and radically separate natures. Visualization is the basis of rational thought and allows for consistency.

Any attempt to define "object", without incorporating shape, will render the assumptions ambiguous and inconsistent because it will allow for multiple "object-based" interpretations, thus destroying any hope of scientific precision. The theorist will inevitably be leaving it up to their audience to pick and choose an interpretation from a host of "object-based" options. Visualization is the "finest" possible tool and will allow one to dismantle any competing definitions by exposing multiple interpretations and counterexamples.

KILL THE OBSERVER

Since the subject matter is existence, we must use reason to isolate the qualities of objects which they have independent of any observers. What properties did the Earth have prior to the evolution of life?

Objectivity is all about "killing the observer." Definitions and scientific claims must be observer independent. The devil's advocate will argue that this is impossible. The process of critical analysis and theory formulation still involves an intellect, an observer. Indeed, it requires a mind to make or understand scientific statements. However, by formulating those claims in terms of qualities which objects can be understood to have independent of observation, one "kills the observer" to the greatest extent possible.

There is a clear distinction between what the intellect can conceive and what is observed through the sensory system. The former is called Hypothesis & Theory, the latter is called Evidence.

The observer who reports what they gather directly through the sensory system is the one that must be killed in science. The "observer" who imagines hypotheses and theories cannot and need not be killed.

It is just imperative that the definitions and claims revolve around an existence which can be understood to persist in the absence of living entities. Alternatively, subjectivity is when one relies on experience garnered through the sensory system.

The concepts of belief, truth, proof, evidence, knowledge and fact rely on sensory input in order to make a determination about whether the statement has been verified or falsified, or judged to be accepted to some degree.

EXISTENCE

I have defined the term "object." I will now proceed to define the crucial term "existence."

In defining existence, is it imperative that shape be incorporated, but this is not the only prerequisite. If it were, then by definition, any object would exist (circles, lines, squares, Superman, Bigfoot, Santa Claus, unicorns, etc.). All of these terms resolve to objects with shape and any that are three dimensional can serve as hypothetical objects within a theory. The reason they do not exist, however, is because they lack the second requirement: location.

In order to have a location, an object must be somewhere with respect to everything else, such as you, me, the Earth, the Sun, the Andromeda galaxy, etc. An object without a location cannot be said to exist.

Location – the set of static distances from one object to all othersDistance – the spatial separation between the surfaces of two objects

Existence – that with shape and location (i.e., object + location)

To exist is to have physical presence; something somewhere. The definition of the terms "object" and "existence" have to be critically reasoned prior to any consideration of the evidence.

Existence is observer independent. Something exists whether there are any apes present to poke at it or not! The Earth, Moon, Sun, Andromeda Galaxy, etc., all existed prior to the evolution of humans. Neither their shape nor their location is dependent upon any observers.

THEORY VS. EVIDENCE

The devil's advocate may concede that the definitions of "object" and "exist" can be critically reasoned, but how do you determine what actually exists? Don't we still have to look at the moon to verify its presence? You can't just pull everything out of your intellect arbitrarily!

First let's define evidence.

Evidence – the body of observations, experiments, measurements, and equations (i.e., anything gleaned through the sensory system).

Evidence will undoubtedly shape hypotheses and theories. I see the moon. I want to formulate a theory to explain what I see.

However, when I do formulate the theory, I start from the beginning — completely divorced from any evidence. I develop my definitions, I conceive of objects with shape, etc. The entire theory can be synthesized and presented without mentioning the evidence. And, before I even consider the evidence, I must make sure I have a rational theory first. Otherwise, I'm engaged in a meaningless effort to compare evidence against something which cannot even be imagined or conceived. Additionally, I recognize that there are always multiple possible other theories which could be devised to explain the same evidence. Evidence never proves theory.

The key is that Theory and Evidence are eternally separate categories. Never the twain shall meet. The former is generated with the intellect from the unlimited possibilities, while the latter is ascertained through the sensory system. While evidence will influence the intellect in the course of developing a theory, when it comes time to present the theory, it can be accomplished without communicating a single shred of evidence. One simply proposes the objects and their behavior along with the definitions.

Once it has been concluded that the theory is rational, science is done. The presentation is over. We can then proceed to bust out all our pieces of evidence and quibble, according to our own personal opinions, about this or that if we'd like. Belief and disbelief, truth and falsity, proof and disproof, evidence for or against, known and unknown, fact or not, right and wrong, correct and incorrect, acceptance and denial, etc., all resolve to opinion. There's nothing wrong with discussing the evidence as long as it is kept in the proper context, i.e., outside of science.

WHERE'S YOUR EVIDENCE?

When you propose a different theory than the one which is widely accepted by scientists, oftentimes, you'll be accused of "ignoring evidence." There seem to be two different notions of evidence which are typically employed, evidence and evidence for.

The former, I have defined above. Any body of accumulating evidence under that definition can have many different possible explanations. Proposing a theory which differs from the accepted one does not mean you are ignoring the evidence. You simply have a different interpretation.

People have been so conditioned to think in terms of belief, truth, proof, evidence, knowledge and fact that they've fallen into the trap of developing dogma with respect to certain theories. To even assert a theory contrary to the widely accepted one is sometimes considered a blasphemy of sorts. This type of thinking relies on the second notion of evidence — evidence for.

Evidence for, is a step beyond mere evidence because it asserts that certain evidence (observations, experiments, measurements, equations) objectively indicates one theory over another. That any given evidence is evidence for a certain theory boils down to someone's personal opinion because it relies on input from the limited sensory system. At best, you could describe it as scientific opinion. The devil's advocate may want to ascribe some "objectivity" to it, but nevertheless it is certainly not on the same plane of objectivity as the proposing of the theory itself.

SCIENTIFIC INSTRUMENTS AND DETECTION

It's important to bear in mind, when discussing physics, that we are dealing with the very fundamental levels of existence. Whether or not we can see or detect something is immaterial. Even in the obvious case of the Moon, the physicist must still ponder the mechanism by which the light reflected from its surface and entered into his retina. His sensory organs are the most basic scientific instruments and he has no choice but to theorize as to how they interact with the environment.

The empiricist will insist, however, that our sensory system is not just limited to our sensory organs; we can use instruments to peer into deep space or the subatomic world. Many times he or she may want to assert that the definition of existence resolves to that which can be detected. When one responds that it would be impossible for distant galaxies well beyond the visible universe to exist, the empiricist retorts that his definition has been misunderstood. He didn't mean detectable by humans, or any other living entity, because of its proximity, he meant that something's existence hinges upon whether it is able to be detected even if no one ever comes along and pokes it. It's an ability that a thing possesses on its own which makes it real.

The empiricist has unwittingly incorporated the critically reasoned definition of existence, i.e., object + location. To say that something can be detected, implies that it has the ability to interact with a detection instrument. In other words, it's an object capable of interacting with another object, and the two are spatially separated until they have surface to surface contact. No matter what detection instruments we use, be they radio telescopes or the Large Hadron Collider, we've already assumed before we even turn them on that they will be interacting with objects in the environment.

Alternatively, the empiricist will try to say that existence means to be measurable or quantifiable. Making measurements and quantifying are empirical processes which add to the evidence. These operations need to be interpreted by proposing a theory as to what was happening physically to account for these observations. Once again, these processes imply interaction between objects which have shape and location.

At the end of the day, the best we can do with these instruments is gather endless amounts of evidence. The readout on the radio telescope is evidence, the readout on the LHC is evidence. This is of course the entire point of empiricism. However, the second we want to explain the evidence, we have no choice but to invoke existence. After all, it was real objects (environment) interacting with real objects (instruments) this entire time. We start at the beginning by conceiving of objects, formulating definitions, etc., and working towards a theory.

Once we have a rational theory, only then can we bring in the voluminous amounts of data and argue according to our own personal opinions whether we think the theory explains the evidence or whether we believe the theory. And, ultimately, we always have to remember that there could be other possible theories of which we have not yet conceived. Evidence never proves theory.

"SUPERMAN EXISTS!"

The devil's advocate will question why we can't just suppose that one of the fantasy entities like Superman exists. Can't we assume that he has a location? Why not?

Let's say he does have a location with respect to the Earth, Sun, Moon, Andromeda Galaxy and everything else. Here, though, the assumptions are open to further analysis. Superman allegedly possesses special abilities, like flying. Without these special abilities he is no longer Superman but just a man. How does he defy gravity in this way? How does he perform this magic trick? The proponent of any such theory will run into insurmountable problems.

A PRIORI PHYSICS

It must be reiterated that thus far no concept of belief, truth, proof, evidence, knowledge, or fact has been incorporated. Only the intellect and not the sensory system has been wielded. The definitions for the terms "object" and "existence" are explicit assumptions whose objectively distinguishing characteristic is that they allow for the greatest possible clarity and consistency. In addition they are understood to be observer independent. Whether or not they're "true" or "correct" is a matter of subjective opinion.

Belief, truth, proof, evidence, knowledge and fact, as illustrated earlier, arise as concepts due to the uncrossable gap between our immediate experience generated by neural activity and "external" existence. We are ultimately ignorant about observer independent existence, and these concepts serve to communicate degrees of certainty, i.e., our opinion, based on sensory inputs.

Thus, the physical sciences are a priori, in the sense that theories are based on explicit assumptions asserted, and analyzed prior to a single shred of evidence.

A theory, being irrational, means that the intellect cannot grasp it or that it is left open to broad interpretation. It is unintelligible, incoherent, inconceivable, unimaginable. It is impossible to wonder whether something incoherent is true or false or whether there's evidence for it.

If I said "blarg exists" would your first reaction be to wonder whether it's true, or whether you believe it, or whether there's evidence for it? No! You'd ask me what "blarg" means. Meaning and understanding always precede any truth considerations because otherwise there's no intelligible claim to assess in the first place!

If someone wants to say that the concepts of belief, truth, proof, evidence for, knowledge or fact are objective they simply need to define these terms such that they can be used consistently. Define truth such that it can distinguish in a black and white manner that which does or does not qualify as truth in every conceivable circumstance. It's impossible. This is precisely why philosophers have debated the nature of truth for thousands of years and continue to do so! And when someone does attempt to define one of these terms they unavoidably end up merely invoking the others in a web of circularity.

CONSCIOUSNESS

Perhaps the devil's advocate will question how the definition of existence can preclude anything perceived within consciousness. After all it feels more "real" than anything else. How can one say that one's consciousness doesn't exist? You are directly aware of it of course it exists!

The devil's advocate simply needs to define the term "existence" and "consciousness" in order to reach such a conclusion. How else would one determine whether consciousness qualifies?

The purpose of a scientific definition is to narrow the usage of a term down to a single interpretation, i.e., to the point where everyone understands the same thing.

Failing to do so inevitably means that the devil's advocate will use the term existence to refer to many different concepts simultaneously. This is precisely how people end up going in circles forever while debating these issues. Without any narrowing of the usage it is unavoidable that everyone in the debate will be understanding things differently, miscommunicating, and chasing their tails.

Consciousness refers to neural activity in the brain. Only objects having location exist, the motion of those objects does not itself exist since motion is not an object. Therefore consciousness does not exist in the strict scientific sense of the term in which it has been defined above. If one tries to conceptualize consciousness as being some kind of "mental existence" you'll run into irrational assertions like a soul or spirit. It won't be possible to understand the meaning of it.

That being said, I do find difficulty in understanding how it is that our sense of subjective experience arises via surface to surface contact between objects in the brain. The question of how the material world gives rise to experience of course has always been a huge question for philosophers, I don't pretend to have an answer. It may be that sheer complexity prevents any such full understanding. Perhaps it's one of life's unanswerable mysteries.

At any rate, full service to these issues would deserve an entire article on its own. Ultimately, what is crucially important is that one can consistently distinguish the concept of existence in science. We are clearly talking about an existence which is independent of any living entity's perception.

A POSTERIORI

The empiricist may try to argue that science cannot be "a priori." When we were born, it was all sensory data that informed us about the world. Everything roots back to empiricism.

Alternatively, they may say that the evolution of life, in general, began with sensory data. The intellect evolved secondarily and was only ever informed by it.

How the present situation came to be is irrelevant for this discussion. Once the ability of the intellect to conceive of the unlimited possibilities has been established, it is no longer bound by the limitations of the sensory organs.

Still, the empiricist may insist that shape and location, etc., are qualities ultimately informed to us by the senses even if we can conceive in an unlimited manner within that framework. Nature may be incomprehensible to us and science must be equipped to deal with that. We can't arbitrarily confine ourselves to what the intellect can or cannot understand.

INCOMPREHENSIBLE NATURE

The proponents of Modern Physics will insist that Mother Nature is just too weird to understand in any way that we're used to. Any time you suggest that their theories are nonsensical this is the immediate response you'll receive. In Quantum especially they are fond of saying that us apes evolved here on the macroscopic scale and so we can't expect our puny primate brains to understand in any normal sense, what goes on at the atomic or subatomic scale. Every single documentary on Quantum prefaces itself by reminding you of this to prepare you for the insanity that you are about to witness.

The problem here is that even if it is the case that Mother Nature is beyond our puny understanding it does not help relieve our ignorance at all to try to fill that void with irrational theories which also cannot be understood. The proponents of Modern Physics are effectively making a sort of God of the Gaps argument. We can't comprehend nature so we must propose theories which we also can't comprehend. When they propose these incoherent explanations, however, we learn nothing. We may as well go back to traditional religion.

They claim, while it's not intuitive, with years of study of the equations, you can gain some kind of "fuzzy" understanding which cannot be communicated clearly to the uninitiated. They have to speak in vague analogies. Only the experts can get some kind of grasp on it and we poor schmucks have to just take them on faith. But it's "reasonable faith" because Modern Physics can make predictions with the mathematics. Of course, the equations still have to be physically interpreted no matter what since they just represent relations between quantities, i.e., numbers. The mathematical physicist wants to deflect being questioned on what they mean by "object" or "exist" by claiming that their understanding of those terms is hidden in the math somewhere.

MATHEMATICS

Mathematics is not the language of physics. Equations do not explain anything. They are purely descriptive. A description of what is measured or observed does not explain the underlying physical phenomena behind what was measured or observed. To be clear, I am not disputing that the equations of mathematical physics can make very accurate predictions. For further arguments on predictions or how technology is developed please see the article mentioned at the beginning.

Equations absolutely depend upon an observer since the variables in the equations represent physical quantities which can be ascertained through empirical investigation. Quantities, i.e., numbers, have to be interpreted, as do entire equations. Take velocity for example.

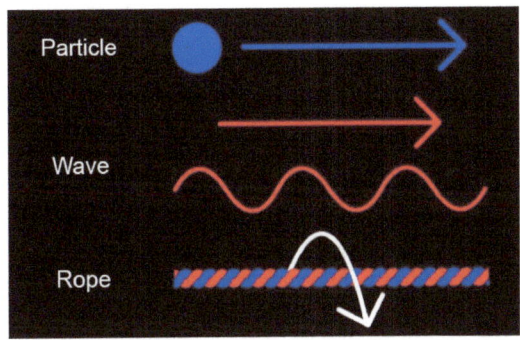

If all you are given is "100 feet per second" you have no idea what the quantity is referencing physically. It could be the motion of a particle, the propagation of a wave, the torsion along a twisted rope, or something else.

All quantities in physics, without exception, are just numbers that need to be interpreted physically. A quantity by itself does not dictate what objects are involved. More often than not, a quantity is a relation between a unit of measurement and that which is being measured. Sometimes they are dimensionless. Either way, without a physical interpretation a quantity is just a meaningless number.

Since all quantities need to be interpreted, and all equations are comprised of relations between quantities, it follows that all equations need to be interpreted as well. Equations by themselves also cannot dictate what objects are involved in reality. They can be formulated from experience without having any understanding of the underlying physical phenomena. An equation is really just an advanced form of pattern recognition.

A classic example of a physicist distinguishing between an equation and an explanation is Newton and his Law of Gravitation. When Newton published it, he reiterated, multiple times, that he did not have a physical explanation for gravity. He was able to discover an equation which described the motion of objects under the influence of gravity, and which could make very accurate predictions, even though he had no understanding, whatsoever, of the mechanism underlying gravitational attraction.

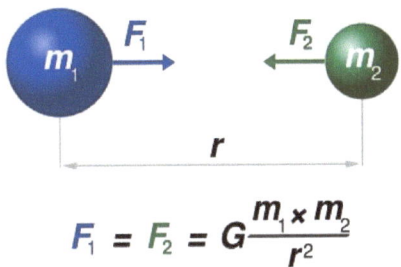

$$F_1 = F_2 = G\frac{m_1 \times m_2}{r^2}$$

In the equation, force is a quantity, big G is a quantity, mass is a quantity, distance is a quantity. The equation tells you how these measured quantities vary with respect to one another and is capable of making very accurate predictions about the observed motion of celestial bodies or bodies falling towards the Earth. Yet, the equation, in no way, acts as some kind of blueprint detailing what is happening in reality that causes these behaviors. It simply relates measured quantities.

All equations in physics deal strictly with quantities which can be ascertained by an observer, i.e., through some kind of empirical means. They say nothing directly about objects or existence. Equations are part of the evidence. All physical interpretations which attempt to explain equations deal strictly with objects and existence.

The proponent of mathematical physics cannot attempt to block criticism of their physical interpretations by asserting that the real meaning of their claims about objects or existence is somehow contained in the mathematics. They absolutely must define these critical terms clearly if their claims are to be understandable at all.

The focus of the CSM on prediction has led empiricists to see equations as the core of physics. That's where the rubber meets the road, because that's how Mother Nature tells us about Herself. Yet, equations do not automatically convey any understanding about objects or causes. Modern Physics has taken the precedence of mathematics to such an extreme that they treat abstract mathematical concepts as if they are real things. Physical interpretations, however, have been relegated to "philosophy" and are considered to have little to no importance.

OBJECT VS CONCEPTS

It's important to distinguish objects from concepts in order to avoid the fallacy of reification.

An object has shape and thus can be visualized or illustrated. This applies whether it's a real standalone entity like the Moon or an abstract object (point, line, square, circle, Superman, unicorns, etc.). Whether it's real, or visualized within the mind, if a term resolves to something that has shape and can be illustrated, then it's referring to an object.

Concepts, however, do not have shape and always invoke two or more objects. Concepts can also become "nested" by relating other concepts which themselves invoke objects. These are abstract concepts and they can continue to become more and more nested into a hierarchy in an unlimited fashion. As a result, concepts cannot be visualized like objects can, they can only be defined and understood.

All terms in a language can be categorized according to whether they resolve to an object or a concept. Whereas concepts are defined and understood, objects are visualized and named.

REIFICATION

The fallacy of reification, also known as the fallacy of misplaced concreteness, is the central problem with mathematical physics. Highly abstract mathematical concepts, such as warped spacetime, black holes, wormholes, singularities, big bangs, wave packets, zero dimensional particles, higher dimensions, forces, fields, charges, energy, etc., are clearly not objects as they cannot be illustrated. They have no shape or location. As such, they cannot be said to exist or perform actions.

Take the field concept. According to Wiki, "In physics, a field is a physical quantity that has a value for each point in space and time."

If I measure around a magnet and record the readings at each location I have a field: It's a list of measured quantities at each location, or a function that represents the same thing.

Of course, this does not tell me what objects are interacting with my instrument (object) to produce the readings. It simply tells me what readings I got. It's a description of what I observed. Mathematical physicists have taken this abstract concept and converted it into an "object" which exists and is capable of influencing other objects such as iron shavings. They have reified the concept. Of course, it's irrational to say that the iron shavings were moved by the "list of measured quantities at each location" or by a function.

This type of sleight of hand serves as the entire foundation of mathematical physics. Their "objects" are always abstract mathematical concepts which they turn around and treat as though they were objects with shape and location capable of performing activities and influencing other objects.

It is clearly impossible to illustrate an abstract mathematical concept. Images of warped space, waves, and magnetic or electric fields, are the reified versions which in no way are meant to be taken literally.

As far as Relativity, Quantum Mechanics, and String Theory go, every single illustration in every physics textbook, every animation in every physics video, every drawing on every board in every physics lecture...out of all of them, not one is meant to be taken literally as an illustration of the shape of the object. This isn't a matter of them having to simplify the drawing — these images in no way represent the shape of any object because they are figurative illustrations of abstract concepts. Even the illustrated particles cannot be taken literally to represent little beads of matter — they have been reified from zero dimensional abstractions!

Modern Physics is all smoke and mirrors.

The endless slew of drawings and animations gives everyone the distinct impression that objects are being discussed and used in theories when in fact they are not.

Mathematical physicists not only draw images, but talk about math concepts like they are objects with shape — "the field vibrated," "the field had an excitation," "the spacetime, warped, curved, bent, expanded, pushed," "the time dilated," "the charge experienced a force in the field," "the particle carried the force," etc.

Despite it being mathematical physics, its practitioners absolutely cannot avoid invoking objects and existence in their presentations.

You cannot do physics without objects. Even if they only use objects figuratively, they still have to imply that objects have shape or else it would become instantly obvious to them and everyone else that their explanations are irrational!

Here let me show you, I've captured some photos of the real "objects" they use:

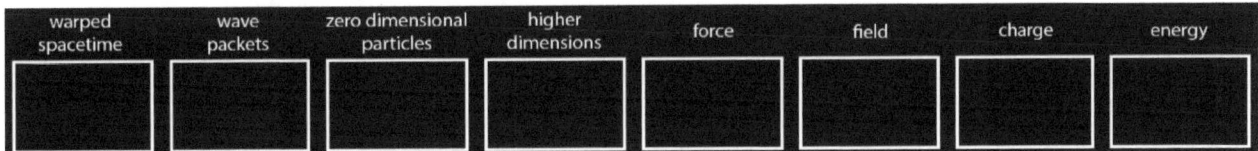

The physical sciences are all about visualization. It's unavoidable. If one cannot conceive of the objects of a theory then one cannot conceive of the explanation and nothing is learned. This is why everyone thinks mathematical physics is so hard to understand — because it's impossible to understand. The mathematical physicists themselves can't understand it. If they ever had to define the key terms "object" and "existence" such that they can be used consistently, the entire house of cards would come crumbling down in an instant.

PARTS AND WHOLES

One common objection to the definition of "object" is the confusion surrounding parts and wholes. A forest is made up of trees but those trees are made of cells which are made of atoms and so on. The devil's advocate might argue that a forest is a concept because it invokes two or more objects.

A forest has shape and thus qualifies as an object. If I start talking about the individual trees then I've simply switched contexts. Now the audience considers the tree to be made of a single piece.

One has to be strict with definitions, but loose with context. If I am presenting a theory and I say "the forest caught on fire," the audience is considering the forest to be a single object for the purpose of understanding my presentation.

Collections of objects are objects (a pile of sand, the atmosphere, a galaxy, etc.) because they have shape. They are bounded and do not blend with their immediate background. One just has to maintain context. When I point to a galaxy and say "Andromeda" I am treating it as a single object.

Parts of objects are also objects. My hand is part of my body and my palm part of my hand. All of them have shape.

Once again we just need to be aware of the context.

DEFINITIONS

I've already provided some definitions and outlined the reasoning behind them but I would like to delve into these issues in more detail.

As stated above, rationality hinges upon clarity and consistency. A scientific definition is one which limits the usage of a term down to a single possible interpretation. This stands in stark contrast to the manner in which definitions are typically conceived.

In science, it is imperative that you define your key terms clearly so that your audience knows what you mean. Otherwise, everyone will walk away with their own interpretation.

In my experience here are some typical criteria proposed for determining the "correct" definition:

1. it's in a dictionary
2. iommon usage
3. experimentally determined
4. authorities say so
5. agreement
6. listing examples

Let me detail each of these individually.

1. Dictionaries only serve to document the common usages of terms. Dictionaries are not meant to convey scientific definitions. There is no focus on maximizing precision but instead on outlining how people typically use the word. This explains why almost all terms in a dictionary have multiple definitions. Dictionaries also make routine use of synonyms and circularity. Someone who just wants to reference a dictionary is trying to shift the burden onto you to determine which definition they mean and how they are using it. This leaves the door wide open so they can shift their arguments later on or accuse you of strawmanning.

2. There is a distinct difference between scientific language and everyday speech. Once again someone who says: "use the meaning that everyone does!" is just trying to shift the burden onto you to figure out what they mean. The same word can be used in many different ways in colloquial speech and there is very little chance that it is typically used with scientific precision.

3. As was dealt with in detail above, scientific definitions cannot be determined through an experiment because they must be critically reasoned prior to the evidence. Any experiment would itself have to be explainable with a theory which utilizes predefined scientific terms.

4. There are no authorities in science. This cannot possibly act as an objective criterion.

5. People may agree to a definition but in no way does this guarantee that the definition will be scientific. The meaning of the term could still be very vague and used in many different ways. Those who agreed to such a definition are bound to miscommunicate and chase their tails.

6. Sometimes an individual claims that all they have to do is list some examples of what qualifies under their term and that is sufficient. They won't even try to define at all. They expect you to abstract away the meaning that they intended from the examples they provided. Once again this shifts the burden onto you and leaves the door open for them to amend their arguments later on or accuse you of strawmanning. Usually, their examples include things which embody different usages of the term in question. This is also

the prime means by which people typically defend terms such as truth or fact. They will list what appear to be very obvious examples and consider the debate settled. What they cannot do is outline specifically what is meant by those terms in order to distinguish that which does or does not qualify in every conceivable circumstance.

Now that I've addressed some common criteria for definitions, I'd like to expand upon the consistency requirement. In order for a definition to be used consistently it must satisfy four criteria, all of which are related. A definition must be unambiguous, non-contradictory, non-synonymous, and non-circular.

NON-CONTRADICTORY

The definition cannot contradict itself. If it does, then there is no possible way to understand what it means or use it consistently. For example, we can't have any zero dimensional objects or square circles.

NON-CIRCULAR

A definition cannot use the term being defined. It also cannot form a chain of circularity. For example, if the definition of term A uses term B, then term B uses C, and then C uses A, you have circularity. Doing any of these fails to narrow the usage at all and leaves the term open to broad interpretation.

For example, we can't say that a line is a row of points and then say that a point is the intersection of two lines. Likewise we can't say a field is made of particles and a particle is just an excitation of a field.

NON-SYNONYMOUS

A definition cannot simply point to another undefined term. Once again, this does not narrow the usage at all, or convey any specific meaning. A common example would be to say that "existence" means "to be" or "real." These do not narrow the usage of the term "existence" at all and, as such, it's left open to broad interpretation.

UNAMBIGUOUS

If the definition satisfies the other three, there's still a chance that it could be left open to interpretation.

The other three are flagrant fouls; ambiguity is more subtle. A definition must be crisp and clear. This means the definition must be able to clearly distinguish, in a black and white manner, that which does or does not qualify. But how is such a determination made? Oftentimes we have to engage in Rational Discourse.

RATIONAL DISCOURSE: "TESTING" A DEFINITION

Oftentimes, someone will employ this process on their own until they're ready to unveil their definition to a broader audience. They will define a term and then try to think of as many examples as possible in order to see if they can easily determine whether the example qualifies or doesn't qualify under the definition. If they discover a counterexample which cannot easily be placed in either camp, then they go back to the drawing board and attempt to refine the definition. They are "testing" the definition against examples.

Once they are satisfied they engage their peers with the definition and challenge them to find a hole in it. This process is crucial to Rational Discourse. Others will then run through examples and try to find one that exposes an ambiguity in the definition. Alternatively, if there's something they don't understand about the definition they will ask the definer to clarify.

Once the definition has been thoroughly vetted, it stands as is until someone can come along and expose a flaw. At this point, the definition should be clear and everyone should be able to understand and apply it consistently.

"WORD GAMES"/"SEMANTICS"/"PHILOSOPHY"

If you ask someone to define their key terms and also "test" the ability of those definitions to be used consistently by trying to produce counterexamples, the person might say you're engaging in "semantics" or "word games" or "philosophy." They are not used to Rational Discourse and are fully unprepared to be challenged. As a way to deflect the onus of defining key terms, they wish to insinuate that the challenger is some kind of word jester delving into irrelevant issues.

There are two different notions at work here. On one hand the accuser is relying on the fact that ultimately a particular word, i.e., noise or combination of letters, is arbitrary. They are implying that your disputation revolves solely around not liking the particular label itself that they are using. This would indeed be word games as the choice of label has no bearing on intended meaning.

The definition associated with that label, however, is crucially important. In order to communicate with another human being that person needs to be able to understand what you are talking about. Otherwise, you both will go in circles for all eternity. Rational Discourse is about exposing misunderstanding, about clarifying and refining.

Almost always, the person who makes such an accusation is themselves the true word jester. They don't define their terms, or define them very vaguely, they use them with extreme inconsistency, they leave it up to their audience to determine what they mean, they switch contexts without warning, they dig their heels in when asked to define or clarify and they all around engage in evasive tactics to hamper any Rational Discourse.

"REDEFINING"

The devil's advocate might say you're just "redefining" a word. They typically mean that you are using a different definition than one which might be used according to the

six unscientific criteria I listed above. They are insinuating that the definition has already somehow been settled and you are just making things up completely out of the blue.

To avoid confusion, it's best to associate a scientific definition with the term it's most closely related to in everyday speech. You probably shouldn't define "object" as "the unbounded formless void" just because you can. Beyond that, however, there is no "redefining," there is only defining. You start with a term and you narrow its usage with a definition.

A LANGUAGE OF CIRCULARITY?

The devil's advocate will argue that all terms are defined with other terms. While the circularity may not be so obvious because of the number of hops required, eventually all words in a language form a web of relations in which they are all codependent. There is no way to avoid circularity.

Of course, this would be impossible because there would be no first point of meaning. If everything was circular, regardless of the number of hops in between, every word would still be meaningless.

This is where objects and visualization come in. Visualization allows you to break out of the circularity. An object is not defined, an object is visualized. Yes the term "object" has to be defined categorically, but objects themselves are visualized. Objects precede all definitions and form the basis of meaning. We don't need to define all terms with other terms, we just need to make sure that the concepts being deployed ultimately root back to relations between objects in a clear manner.

INFINITE REGRESS

The devil's advocate will now make a similar argument and claim that one could endlessly ask what something means, and once given the definition, ask what the terms in that definition mean, and so on and so forth. As per the previous section once we reach objects they are not defined but visualized. It makes no sense to point at an object and ask, "what does this mean?" It just is and you visualize it. It cannot be reduced any further.

That's not the only form of infinite regress, however. Oftentimes, inquiring too far about how something works can get you accused of asking "why" or "how" to an unreasonable extent. They'll say, at some point, we have to accept that we won't be able to understand any further.

Proponents of Modern Physics will often levy the infinite regress accusation when they are questioned. Here's a video they love to post of famous physicist Richard Feynman using this line of thinking to deflect a basic question about how it is that magnets attract:

https://www.youtube.com/watch?time_continue=6&v=36GT2zl8IVA

Once again, we terminate at the object level. Once the theorist has provided their fundamental objects we can go no further. Objects themselves cannot be explained, they are the ultimate given. So long as the mathematical physicist speaks in terms of abstract mathematical concepts like "forces" it's perfectly fair to ask them endless questions. Feynman clearly has no clue why magnets attract. He's trying to insinuate that it's some kind of incomprehensible mystery to some extent which we just have to accept rather than being a matter of his own lack of understanding.

There is one last notion of infinite regress that I can identify which is also related to the others. Once we hit the object level someone may be tempted to ask what the object is made out of. It's a sort of metaphysical question, a question about the ultimate nature of "reality itself." What is it that actually fills the shape out? What is the substance, what is the matter?

It is impossible however to reduce any further. Shape is the only inherent quality of an object. In what form could the answer to such a question ever arrive? Inevitably, any such answer will always be a concept which itself will be able to be reduced to relations between objects.

DEFINITION COMPETITION

At the end of all this, the unrelenting devil's advocate may argue that such lofty requirements of definitions are tantamount to demanding perfection. There is no such thing as absolute consistency or the complete elimination of the possibility of miscommunication. Single interpretation precision is impossible.

Of course, "perfection" and "absolute" are opinionated terms. It's arguable that there's always some degree of ambiguity or potential for miscommunication no matter how slight. The point of Rational Discourse and scientific definitions however is to bring about as much clarity and consistency as is humanly possible even if it's not "perfect."

Typically the devil's advocate makes this argument because they want to go right back to vague definitions and inconsistency. They want to pretend that precision is an unrealizable ideal and as such they are not required to clearly define their terms. Regardless of whether "perfect" precision is possible, the devil's advocate has to compete with the other definitions that are out there for title of most clear and consistent.

As explained earlier, an object lies in stark contrast to its immediate background. They are of clear and radically separate natures. Definitions for concepts which don't clearly root back to objects, will always lose in the competition as multiple possible interpretations will be able to be exposed.

Knowledge, fact and prediction are not formally incorporated into the RSM.

Additionally, theories never become facts. As stated earlier — belief, truth, proof, evidence,

HYPOTHESIS

1. exhibits (objects)

2. definitions for Key Terms

3. statement of the Facts (paint the initial scene)

First, the theorist presents to his audience the objects of his hypothesis. Since we have already reasoned that all objects have shape as their only inherent quality, he must illustrate those objects. He can also bring in 3D models.

Given that oftentimes the subject matter will be complicated, he can simplify the drawings or models if need be. If we're presenting the Rope Hypothesis version of the atom, for example, we don't need to try to draw every rope as obviously that would be impossible.

Second, he asserts the definitions for his key terms, the ones that will make or break his theory. The definitions must be clear and able to be used consistently.

Third, he demonstrates the statement of the facts. He paints the initial scene of the movie by illustrating how the objects of his exhibits are arranged with respect to one another. To be clear, this is his statement of the facts, they are not facts in an of themselves. These are assumptions he is making about how things really were in reality.

These are the three stages that one must traverse in order to critically reason and present a rational theory. As mentioned earlier, of course, evidence will influence how one develops a theory. However, when it comes time to engage in the scientific method and present the theory, it must be built from the ground up. And nowhere in the RSM is there any provision for evidence whatsoever.

The conclusion is predicated on the theory. The theory is predicated on the hypothesis. Without Hypothesis there can be no Theory, without Theory there can be no Conclusion.

To put it in the most succinct possible manner, the RSM is a movie. The hypothesis is the first frame. The theory is the rest of the reel. The conclusion is where you determine whether or not it is rational.

THE RATIONAL SCIENTIFIC METHOD

The RSM is broken down into three distinct steps — Hypothesis, Theory, and Conclusion.

In the CSM, hypotheses can become theories if enough experts raise their hands and decide that there's enough evidence. There's no objective threshold which distinguishes them. The opinionated matter has no choice but to come down to a vote. If the experts decide that they have even more confidence in a theory, then it can be converted into a fact.

TO RECAP

The visualization approach accomplishes the following:

1.   provides a basis for clear understanding of an explanation for a consummated event

2.   is the most objective possible form of communication (any miscommunication can be ironed out)

3.   allows for terms like "object" and "existence" to be defined in an observer independent manner

4.   allows one to break out of circularity and infinite regress

5.   allows for the consistent formulation of definitions and usage of key terms

6.   allows for the exclusion of opinionated concepts like belief, truth, proof, evidence, knowledge, and fact.

In the RSM, hypotheses and theories are distinct stages of the method, not separate classes of proposals altogether. Hypotheses and theories are eternally separate. Hypotheses never become theories. The difference between them is objective. You don't need to look at "scientific" consensus to determine whether something is a hypothesis or a theory.

THEORY

1. present a movie of the behavior of the objects

2. explain the behavior

First, the theorist presents a movie of the behavior of the objects. Whereas the hypothesis is a static freeze frame, the theory is dynamic. This is the objective difference between hypotheses and theories.

Second, he explains the behavior of the objects. The behavior should follow from the shape of the objects so everyone can understand why the objects do what they do.

CONCLUSION

Each member of the audience decides whether the theory is rational. Rational Discourse regarding the theory can help break down the definitions, analyze the objects,, and discuss their behavior. The theory is either rational or irrational.

RATIONALITY REQUIREMENTS

There are four requirements for the theory to be rational:

1. The objects must be able to be illustrated (at least in simplified form).

2. The key terms must be defined such that they can be used consistently (and must actually be used consistently).

3. Every process referenced in the theory must be associated with a movie that is not missing any frames (at least in simplified form).

4. The behavior of the objects must be able to be understood as following from their shape.

The first three are pretty straight forward. The last one requires elaboration. If the objects of your theory behave in random, sporadic ways for no identifiable reason then we are left with no understandable explanation for a phenomenon. The behavior should be able to be understood as resulting from the shape of the objects. This would exclude things like action at a distance. One particle cannot act on another without coming into contact with it or without some kind of physical intermediary.

SCIENCE

Once again, the empiricist may be tempted to say that the above requirements are arbitrary. How can we demand that nature follow our rules?

We can't, and we are not! We are attempting to achieve a rational understanding of nature. These are the elements that are absolutely required in order to accomplish that. Once we have a rational theory, we can finally say that we have an understanding of one possibility. Mother Nature may not care about what the apes demand of Her. She may be incomprehensible to us. However, without these criteria being satisfied the ape will inevitably be left in a state of confusion and ignorance. It does not help this situation to flaunt the requirements by trying to supplant our ignorance with incomprehensible theories which we also cannot understand. And, no amount of empirical investigation or evidence gathering can ever change that, because, at some point, we have to hypothesize, theorize and conclude.

This is the essence of science; rational explanations! This is how we distinguish the scientific from the unscientific. It's not through belief, truth, proof, evidence, knowledge, fact or any other subjective criteria. It's by using our intellects, critical reasoning skills and rational understanding.

VISUALIZATION

Visualization is the universal human language. It's a capacity which we all possess within our intellects. It allows for understanding, and, between humans, for the possibility of objective communication. Any confusions can be elucidated and driven out to the greatest extent possible.

Every individual is in a position to "verify" that the theory is rational for themselves. There is no need to depend upon external authority. Whether the objects can be visualized, the concepts understood, the key terms used consistently, etc., is determined within the individual using a common human capacity.

An analogy to bitcoin could be made. Each individual node has the entire block chain and can verify it. It's a trustless system. Likewise, each person in science has the entire picture and can understand it. The RSM is the trustless system of science. We don't need to take the experts on faith and we don't need to trust that they haven't manipulated the data. Authority and faith cannot possibly survive.

Visualization is so central to the RSM that it helps define the method itself. It's what allows for the objective delineation between Hypothesis and Theory, as the concept of motion is predicated on objects. Unlike the CSM, where the difference boils down to opinion, once again, in the RSM there is no need to ask the experts what stage of the Scientific Method we are discussing.

And, with objects as the core, an entire set of definitions can be teased out through critical thinking. Some are alarmed at how proponents of the RSM insist on the same set of definitions. They view it as a sort of dogma all its own. Yet, the definitions are not arbitrary. They are rooted in objects and can be used consistently. They constitute THE scientific definitions of Physics. Anyone is free in Rational Discourse to expose flaws or produce a competing set of their own. Proponents of the RSM did not invent visualization. This is open source software that could very well run on intellects around the Universe! But, the devil's advocate will claim that visualization is not universal. What about people who were born blind? Are they barred from science?

(Read the amazing story about Esref Armagan, the artist with no eyes.)

Here is a painter who was born blind. He ascertains the form of objects through touch. They even bring him to a historic building in Florence, whose complex geometry makes the perspective extremely difficult to draw, even if you can see. After feeling the building, and also a model of it, he is able to draw the perspective correctly. Even without sight, he can understand the concept of boundedness, the inside from the outside, the object with shape and the unbounded space that it has as its background. I guarantee he could do physics!

THE CSM

The CSM is very practical. When I criticize it I don't mean to say we should never engage in it. Experimentation and pattern recognition are extremely useful for various ends such as the development of new technologies. The RSM, on the other hand, does not necessarily have any practical purpose at all. It's simply a formal means of proposing explanations for phenomena.

While the CSM can be very useful, it can also be dangerous. The focus on data, statistics, correlations and mathematical modeling leaves a great deal of room for confirmation bias, data dredging, and deceit. One website tracks how often papers are retracted for various reasons including deliberate manipulation of data and fraud.

The focus on data and mathematics is very useful for those who want to use the label of science to push an agenda, such as governments or government connected corporations. Even absent these corrupting influences, we have to remember that scientists are people too — the temptation to use peer review as a system of control to ostracize dissenters and protect cherished and entrenched theories no matter how flawed they are is huge. There is no formal conspiracy required, one just has to look at the incentives at work.

The type of environment that has been created out of this is one of group think and ridicule. The notion that science is an open affair where all ideas are consideredm and where the CSM "keeps them honest" despite all the flaws of humans beingsm is extremely naive. What we have are witch huntsm where dissenters are labeled "denialists", and where so called "skeptics" are no more than gatekeepers for entrenched theories. Their only claim to skepticism is when it is applied in extreme fashion towards anyone who promotes an alternative theory.

In the RSM, there is no acceptance or denial. Skepticism, if it means anything, is automatic within the RSM as there is no provision for belief. The RSM is the perfect synthesis of open mindedness and exclusion — one can simultaneously consider many alternative explanations for the same evidence. There is no dogmatic proven truth, and yet, there is still a narrowing process because the intellect is used to eliminate irrational theories.

The RSM can act as a bulwark against the negative effects of the CSM and help to make us more impervious to those, especially in governments, who would want to use the label of science to enhance their power at the expense of our individual liberties.

FINAL THOUGHTS

Of course, I don't hold that everything in this article is perfect. It's also not comprehensive — there's much more that can be discussed. The goal is to introduce the RSM and get the conversation going. I'm sure you can pick holes in it. There's always room for further analysis.

While the RSM formally identifies a consecutive and multi-stage process, realistically, Rational Discourse never ends. We constantly go back to the basics and reevaluate our assumptions. We find the tiniest nuances and bring them out in the open for discussion. Everything is tentative, we never really settle on anything completely. We narrow, we refine, we expose, we ponder. We illustrate, we define, we explain!

GLOSSARY

Angle – the pattern formed by two straight edges fused at one end of an object lying on a plane facing the observer

Assumption – a statement taken at face value for the purpose of understanding a theory

Concept – a term which invokes two objects or two locations of an object (relationship between objects or, nested concepts)

Conclusion – the stage of the Scientific Method where a theory is determined to be rational or irrational

Continuous – not having discrete parts, indivisible

Definition – a limitation placed on the utility of a word

Dimension – one of three mutually perpendicular directions in which an object may face (i.e., length, width, and height)

Distance – the spatial separation between the surfaces of two objects

Energy – the ability to do work

Edge – the attribute of an object that provides closure and distinguishes it from the background

Evidence – the body of observations, experiments, measurements, and equations (i.e., anything gleaned through the sensory system)

Existence – that with shape and location (object + location)

Force – one object's push or pull upon another

Hypothesis – the stage of the Scientific Method where one provides the exhibits, definitions for key terms, and statement of the facts

Length – the continuous matter between two surfaces

Drink Rationally!

Line – an elongated rectangle

Location – the set of static distances from one object to all others

Motion – two or more locations of an object

Object – that which has shape

Perpendicular – a pattern formed by two straight edges: one lying horizontal and the other standing vertical

Physics – the study of existing objects (typically dealing with fundamental and low level composite objects)

Physical – an adjective which qualifies an object as being three dimensional

Point – a small circular dot

Rational – describes any communication in which the key terms have been defined to the point where there is only a single possible interpretation for them (i.e., where everyone understands the same thing)

Science – the body of communication pertaining to the Scientific Method

Space – shapelessness (i.e. vacuum, emptiness, nothingness, formlessness, the void)

Theory – the stage of the Scientific Method where one illustrates the behavior of the objects in the hypothesis; an explanation of consummated events

Time – a comparison between two motions

Universe – all existing objects and space

Words Mean Things

By MonkEmind

Definition and Context are Important!

"It depends on what the meaning of the word "is" is." – President Clinton

At the "higher" level of government, politicians understand full well the importance of defining one's Key Terms. Take for example the term "coup." One would think that it would be a simple matter of determining whether or not what is going on in Egypt (2013) is related to a coup, if the word is defined.

The most common meaning of coup being the following:

Coup: (also coup d'état) a sudden, violent, and illegal seizure of power from a government: he was overthrown in an army coup

http://oxforddictionaries.com/us/definition/american_english/coup

It is obvious to me that U.S. officials ARE familiar with the definition, and also why they would wish to have ambiguous definitions for THEIR key terms.

With accusations that Israel is behind the military coup in Egypt, and the amount of money promised (1 billion) at stake, it is no wonder US politicians want to consider re-defining the word 'coup.

"'If Morsy's removal were to be called a coup, under U.S. law, more than $1 billion in military aid to Egypt would have to be slashed."

"Senior U.S. officials say the administration is examining three potential options – calling events in Egypt a coup and cutting off aid; calling it a coup and issuing a national security waiver; or not determining it a coup, recognizing that the military has taken steps to move the country toward a civilian transitional government and move toward elections."

"So our decisions with regards to the events that have happened recently in Egypt will be - and how we label them and analyze them will be made with our policy objectives in mind, in accordance with the law and in accordance with any consultation with Congress," he (Carney) said."

http://security.blogs.cnn.com/2013/07/08/u-s-avoids-calling-egypts-uprising-a-coup/

There are also political motivating factors in defining, or shall I say- NOT defining, or defining ambiguously, certain terms in science.

If physicists were forced to define in unambiguous terms their magic words energy, field, space and time, their whole house of cards would collapse, and money for projects like CERN and Fermilab would dry up over night. Careers would come to a standstill, funds would dry up and labs would be shut down around the world.

Besides the obvious political motivation(s) behind defining terms, or using "labels," as our friend Mr. Carney called it, there are other (apparently) less obvious reasons for our particular use of words. One may be predisposed to thinking about certain things in certain ways depending on their language and their culture. Who would of thunk it?

From Psychology Today: How Culture Shapes Thought by Lawrence T. White, Ph.D., and Steven B. Jackson

"The evidence is clear: To a surprising degree, language and culture influence how we think about time."

http://www.psychologytoday.com/blog/culture-conscious/201201/do-cultures-segment-time-differently

Wow! Startling! Who would have guessed? A number of "scientific" studies by psychologists, linguists and anthropologists have come to this startling conclusion based upon a number of recent experiments.

Let's look briefly at some of the evidence and conclusions about how and why persons use the words they do for time and space. We'll draw from two seminal articles on the subject:

Does Language Shape Thought?: Mandarin and English Speakers' Conceptions of Time by Lera Boroditsky

And

The Thaayorre Think of Time Like They Talk of Space by Alice Gaby

Boroditsky notes that English speaking, and Mandarin speaking individuals talk about time differently. Whereas English speaking folks refer to time as "moving" from left to right, Mandarin refers to time as "moving" from up to down. In other words, if asked to place January and February in order, English speaking individuals place January to the left of February and Mandarin speaking individuals place January above February.

Of course, everyone notices that English reads and writes from left to right and Mandarin from top to bottom. The experiments looked at both monolinguals and bilinguals. It was concluded:

"(1) language is a powerful tool in shaping thought about abstract domains and (2) one's native language plays an important role in shaping habitual thought (e.g., how one tends to think about time) but does not entirely determine one's thinking in the strong Whorfian sense.

Key Words: Whorf; time; language; metaphor; Mandarin."

Note: the Key words listed were not defined, apparently just used to draw Google hits.

That article asks questions, such as; Do different people who speak different languages look at the world differently? Do people who speak more than one language think differently when speaking a different language? Is thought determined entirely by language?

Boroditsky does acknowledge the inherent difficulties related to the experiments. Of course, I consider experimentation extra scientific, so will not focus primarily on the experiments themselves, and data analysis, etc. My intention is to merely mention the particular ways in which it has been observed that different language speaking persons talk about space and time and the resulting conclusions arrived at by the experimenters.

According to the studies, experiences about the phenomena of time inform individuals that any particular event happens only once (sorry no Ground hog Day) are unidirectional, and this appears to be universal across differing languages and cultures.

"However, there are many aspects of our concept of time that are not observable in the world. For example, does time move horizontally or vertically? Does it move forward or back, left or right, up or down? Does it move past us, or do we move through it? All of these aspects are left unspecified in our experience with the world. They are, however, specified in our language—most often through spatial metaphors. Across languages people use spatial metaphors to talk about time. Whether they are looking forward to a brighter tomorrow, proposing theories ahead of their time, or falling behind schedule, they rely on terms from the domain of space to talk about time (Clark, 1973; Lehrer, 1990; Traugott, 1978). Those aspects of time that are not constrained by our physical experience with time are free to vary across languages and our conceptions of them may be shaped by the way we choose to talk about them. This article focuses on one such aspect of time and examines whether different ways of talking about time lead to different ways of thinking about it."

If the experimenters would have defined their Key Terms like time unambiguously they would understand how ridiculous those questions are. Those "ASPECTS" are unspecified in our experience but are specified in our language precisely because time has not been defined scientifically, explained, or understood.

Different ways of talking about time naturally lead to different ways of thinking about it. Defining terms is paramount in understanding what people are talking about.

English speakers use "front/back words" to refer to time such as before, after, behind, and ahead; mostly ordering events by using the same words as those used for describing HORIZONTAL SPATIAL RELATIONS."

Mandarin speakers also use spatial relations to refer to time, such as front and back, but they also use vertical metaphors, such as up and down (used less often in English).

Boroditsky asks: "Does Metaphor Use Have Long-Term Implications for Processing?"

The conclusions based on the experiments answered her questions:

"As predicted, English speakers answered purely temporal questions faster after horizontal primes than after vertical primes.

"When answering questions phrased in purely temporal earlier/later terms, Mandarin speakers were faster after vertical primes than after horizontal primes. This pattern was predicted by the fact that in Mandarin vertical metaphors are often used to talk about time."

Did we really need to do experiments, along with their descriptive statistics and analysis, to arrive at that conclusion? How we think affects our speaking and how we speak affects our thinking. What is so mysterious about this?

How we habitually use words obviously affects the way we think. We didn't need experiments to understand this, we simply observe it regularly in our every day lives, and can conceive that this is the case with ALL words, not just space and time.

Our observations confirm our observations. Great! But how do we explain this? Words mean things! How we use words in conversational language is different than how we use our words in science, but it is ALWAYS important to understand the words we use.

Let's take a look at Gaby's article. We note that regardless of language or culture humans tend to speak about time in terms of space.

It's no wonder relativists believe that time combines with space because they think space is a substance. It is no wonder that relativists think of space as a substance because they reify concepts into objects, that is, they turn concepts (verbs) into the nouns of reality. Why? Because they were never taught to define terms unambiguously, and in non-circular, non-synonymous and non-contradictory ways. They were never taught NOT to reify, and they never learned to use their words consistently in their presentations. In short, they don't understand the difference between objects and concepts!

Let's continue. Gaby compares an Australian aboriginal group (both monolingual and bilingual) with English monolinguals to conclude the same thing that Boroditsky did;

"The way people conceptualize time is shaped by a range of external influences, both linguistic and non-linguistic."

Other researchers determined the same thing as well. Hebrews order time from right to left, The Kuuk Thaayore from east to west, the Mandarin from up to down (or away to towards) and Tagbanwa from bottom to top. All this time/space mapping is according to a relative frame of reference (damn those relativists...they're everywhere!).

The interesting thing about the Kuuk Thaayorre is their mapping of time and space using an "ABSOLUTE DESCRIPTION OF SPATIAL RELATIONSHIPS." They always map time from east to west no matter which way they are facing. If talking about a long time ago, they point to the east, for instance.

Their north and south frame of reference is anchored to their coast line.

"The terms -kaw "east" and -kuw "west" are defined by the sun's trajectory, while the terms roughly translated as "~north" and "~south" (-ungkarr and -iparr, respectively), more accurately align with an axis defined by the local coastline, forming an axis rotated almost 45° clockwise from that perpendicular to east-west."

They keep track of time using the sun and the moon, and also seasonal variations in flora and fauna. It is very interesting how they use sand paintings when telling stories. For instance, they may represent something to the east as in the past. They may erase something and draw over it for events that took place concurrently.

Another article of note by Boroditsky is entitled: How Language Shapes Thought, which has this to say:

"Language also appears to be involved in many more aspects of our mental lives than scientists had previously supposed. People rely on language even when doing simple things like distinguishing patches of color, counting dots on a screen or orienting in a small room…"

But this is the more salient point, I think:

"Speakers of different languages also differ in how they describe events and, as a result, how well they can remember who did what. All events, even split-second accidents, are complicated and require us to construe and interpret what happened. Take, for example, former vice president Dick Cheney's quail-hunting accident, in which he accidentally shot Harry Whittington. One could say that "Cheney shot Whittington" (wherein Cheney is the direct cause), or "Whittington got shot by Cheney" (distancing Cheney from the outcome), or "Whittington got peppered pretty good" (leaving Cheney out altogether)."

Don't be fooled by the politician's or scientist's word magic! Always make them define their Key Terms.

Not only is it important to define one's key terms, it is also important to understand the context in which they are being used.

To find out more about why true/false, right/wrong, proof, belief, evidence, and authority are NOT part of the Rational Scientific Method of inquiry, read the entire series of Rational Science books by Monk E. Mind. Available on Amazon in Paperback and Ebook as well as Audible and iTune formats.

Or, join other rational scientists at the facebook group: Rational Scientific Method

We have a wealth of resources from books, and articles, to blogs and YouTube videos.

Wherever your interests lie, having a good foundation in the scientific method of inquiry will allow you to decide for yourself what is possible or NOT possible.

You'll be able to spot nonsense from a mile away, and save yourself the trouble and expense of reading mainstream scientific books of impossible and magical nonsense

Special Relativity made simple is the ideal book for laymen with open minds who suspect the fantastic explanations they hear and read about from authoritative 'scientific' sources. Does it make sense to say that twin brothers could differ in their ages by 50 years? Is it rational to say that you can travel to the past or to the future?

Special Relativity made simple addresses these and similar questions and arms the average reader with arguments that enables them to challenge Einstein's theories on the Internet and in conferences.

To obtain a paperback, Paypal to bill@youstupidrelativist.com
USA/Canada   US $20.00 Plus $10 Shipping
Europe 20.00 Euros Plus 10 Euros Shipping

A HISTORY OF SCIENCE... AS DECIDED BY PEER REVIEW

"Sorry Pythagoras, all of the evidence available shows the Earth is, in fact, flat."

"Sorry Galileo, all of the evidence available shows that the Earth is the center of the Universe."

"BTW we're gonna murder you, LOL"

"Sorry Bill Gaede, the evidence available proves that all phenomena of reality is mediated by abstract mathematical concepts. Take a math class!"

MISSING

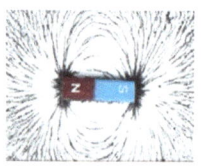
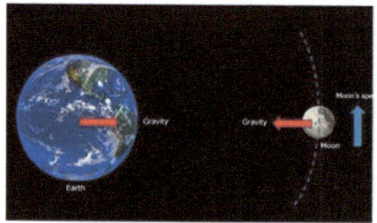

HAVE YOU CONCEIVED OF THIS OBJECT?

Physical Mediator

Light - Gravity - Magnetism - Electricity

Last assumed to be formless abstract concepts like fields, forces, waves, energy, charges, particles, etc. After a century of trying to prove that phenomena of reality is mediated by abstractions, we are turning our search to physical objects whose architecture can justify the behavior mentioned above.

If you have any information please join the Rational Scientific Method group on facebook at https://www.facebook.com/groups/RationalScientificMethod/

PLEASE - INFORMATION NEEDED

Mike Huttner

Art by Daniel Ferguson

Rope Hypothesis
and
Thread Theory

By
Monk E. Mind

www.ingramcontent.com/pod-product-compliance
Lightning Source LLC
Chambersburg PA
CBHW051211220526
45473CB00003B/983